WRECKED

WRECKED

The *Edmund Fitzgerald* and the Sinking of the American Economy

Thomas M. Nelson with Jerald Podair

MICHIGAN STATE UNIVERSITY PRESS | *East Lansing*

Michigan State University Press
East Lansing, Michigan 48823-5245

Library of Congress Cataloging-in-Publication Data
Names: Nelson, Thomas M., author. | Podair, Jerald E., 1953– author.
Title: Wrecked : the Edmund Fitzgerald and the sinking of the American economy /
Thomas M. Nelson with Jerald Podair.
Description: East Lansing : Michigan State University Press, 2025. |
Includes bibliographical references and index.
Identifiers: LCCN 2025005380 | ISBN 9781611865417 (paperback) |
ISBN 9781609177874 | ISBN 9781628955507
Subjects: LCSH: Edmund Fitzgerald (Ship). | Shipwrecks—Superior, Lake. |
Manufacturing industries—United States—History—20th century. |
Industrial policy—United States—History—20th century. |
United States—Economic conditions—1971–1981. |
United States—Economic conditions—1981–2001.
Classification: LCC G530.E26 N45 2025 |
DDC 917.74/90443—dc23/eng/20250325
LC record available at https://lccn.loc.gov/2025005380

Cover design by Erin Kirk.
The launch of the *Edmund Fitzgerald* at the Great Lakes Engineering Works at River Rouge, MI, on June 7, 1958, used with permission from the Great Lakes Shipwreck Museum.

Visit Michigan State University Press at *www.msupress.org*

To the SS *William Clay Ford*, you didn't have to be asked twice.

"A ship in a harbor is safe but that is not what ships are built for."

—JOHN A. SHEDD

Contents

Preface

MY UNCLE DAN WAS THE COOL UNCLE. RIGHT BEFORE I ROLLED OVER INTO DOUBLE digits, I'd spend a week or two in the summers in the mid-1980s at his bachelor pad in Northfield, Minnesota, while my dad, the Lutheran pastor, attended workshops at St. Olaf College across town. So, when my dad said Dan was coming over for a motorcycle trip, I couldn't wait.

We set off from Little Chute, Wisconsin, for the Upper Peninsula of Michigan in late August. My hometown sat an hour south of the state line, just far enough for this ride to count as a road trip.

Summer was turning into fall and leaves paled yellow on mighty oaks lining Highway 41, the country's north–south version of Route 66. As my father zipped down a ribbon of black, freshly laid asphalt, matted leaves spun up around me on the back of his 1980 Yamaha XS. I held onto those last days of summer as tightly as I hugged my dad. I didn't want to fall off the bike and I didn't want the trip to end. Ever.

Before we hooked around the curve of the UP—the bend in the peninsula's elbow—and into lower Michigan we overnighted near Sault Ste Marie. On our family camping trips it was tradition to find the nearest KOA campground because they usually had swimming pools, the treat for well-behaved kids who didn't fight in the backseat of the car—or in this case monkey around on the back of a motorcycle and tumble out onto the pockmarked, weather-worn roads. I was disappointed this time because there were no such campgrounds in the area. We would have to settle for a local mom-and-pop.

Our campsite was located on the shore of Lake Huron and along a vacant, unkept beach. It checked the box for our requirements—a body of water in which to take a dip and cool off from a long day on the road. It would do just fine.

As I waded into the shallow brown waters, mighty ore ships passed by carefully and quietly in the distance.

I could see them.

And I could feel them.

As the ships motored by, their mighty bows pushed cold lake water toward the shore, which became ripples, which became waves, sending my frail eight-year-old body tumbling backward. And when the boats were just dots on the horizon, a tide sucked me out into the surf. What took the moon half a day to do only took these ships fifteen minutes.

I never escaped the magnetic pull of those memories. It is a big reason why I am writing this book.

When the movie *The Perfect Storm* came out a couple decades later, I blew my top. The promotional poster of that little fishing boat climbing a massive one-hundred-foot wave seemed ridiculous at best, sacrilegious at worst. If any film of bravery and heroism on the high seas should be made, it shouldn't be about the *Andrea Gail*. It should be about the *Edmund Fitzgerald*. But it wasn't.

If you are from the Midwest, and especially Wisconsin or Michigan, and especially if you were a child of the 1970s or 1980s, you learned early on about the story of the wreck of the *Edmund Fitzgerald*; the tale of brave sailors who went into the great unknown, never to return. It is our Lexington Battle Green, a significant moment in history that we can call our own.

But instead of a salute to a mighty ore ship that built the American economy of the mid-twentieth century, we were treated to a two-hour romp with a pair of knucklehead sex symbols who irresponsibly barreled into a storm of the century knowing full well that the seas would capsize their little tugboat and kill the entire crew. But they went ahead anyway, because at stake was a once-in-a-lifetime catch that would make them rich. Rich! ("*The Perfect Storm* has all the impact of watching a friend play Nintendo," read one review at the time.)

Rather, it was the Fitz—as the *Edmund Fitzgerald* was affectionally and ubiquitously known—that was the true profile in courage. She went up against *the* perfect storm, and in more ways than one. Yes, she faced a storm of the century (unbeknownst to her crew), but the crew faced a series of challenges: a vessel poorly maintained due to mismanagement, insufficient communication equipment, no depth finders, antiquated meteorological systems, broken weather beacons, downed radar, and a cargo hold loaded to the gills with the season's last haul of taconite ore. ("[Captain Ernest] McSorley never saw waves that high because he never sat that low in seas!" his protégé and former Great Lakes sailor Bob Hom told me, reflecting on the ship captain's last radio communications.[1])

The Fitz wasn't just any old ship; she served as a vehicle—literally and figuratively—that drove the American manufacturing economy of that era and supported millions of jobs and working families all throughout the Midwest and beyond. The ship's namesake and owner, Edmund Fitzgerald, president and CEO of Northwestern Mutual understood this unique, expansive economic ecosystem well. In the wake of the ship's sinking, *Milwaukee Sentinel* reporter Carl Mueller reflected on this important point: "Fitzgerald recognized that the Great Lakes provided a vital, low-cost water highway between the nation's coal, ore and grain producing heartland and industrial centers in the East."[2]

Taconite ore mined in Minnesota's Iron Range was shipped by the likes of the *Edmund Fitzgerald* from the docks of Duluth, Minnesota, and Superior, Wisconsin, to the great steel mills of the Upper Midwest, which would be used to make cars in Detroit or machine tools in Milwaukee. The ships themselves were built in Toledo, Detroit, and Cleveland, and construction was financed by banks and insurance companies in Milwaukee and Chicago.

How the brave crew of the *Edmund Fitzgerald* handled themselves professionally, emotionally, and psychologically over the next twenty-four hours of the voyage would have implications that would go well beyond their own fates. Of course, in the end, the elements were too much, and the catastrophic, corrupt mismanagement of Oglebay Norton Corporation—and utter unprofessional and irresponsible behavior of the ship's owner, Northwestern Mutual of Milwaukee were too egregious. Those brave twenty-nine crew members left behind scores of family, neighbors, drinking buddies, fellow sailors, and friends along the north shore of Wisconsin and all the way to the ports of Ohio. To this day that fateful voyage of the Fitzgerald continues to haunt the widows, the sons, and the daughters of the fallen. Some cannot bring themselves to talk about it, as I learned when knocking on doors in Iron River, Moquah, Sturgeon Bay, and Ashland, Wisconsin. On the other hand, some were more than happy to chat. "You know, I don't get a chance to talk about this. Everyone else is gone," said Bob Hom. "I didn't think anyone cared about the story anymore," said James Haskell, brother of second assistant engineer Russell Haskell.[3]

Personal stories still remain in the recesses of relatives' minds, memories that are well kept but untold. "You must get these kind of requests, interviews all the time," I asked Fred Armagost, the little brother of Third Mate Michael Armagost, outside his log cabin in the woods near Iron River. "Nope. This is the first one," he responded. For Fred, the next hour was hard—but it was also cathartic. "He was a good brother," he said no fewer than five times.[4]

Behind the faces and the names of the twenty-nine sailors were a nation of faceless, nameless paper makers, loggers, miners, mechanics, cooks, waiters, automakers, steelmakers, and teachers, just struggling to make ends meet, who went to work on their own version of the Fitzgerald—a mine, paper mill, assembly line, classroom, or street corner diner. While most were not taking life-or-death risks, not a few were up against their own "bad bosses"—multinational corporations totally unprepared to handle a changing economy but totally willing to cut jobs to fix *their* problems. Indeed, in a four-month period in 1982 the United Steelworkers lost more than 250,000 dues-paying members—a crisis that hit every steel town in the country.[5]

Businesses looked the other way when workplaces became rundown and unsafe, when technology investment went to the non-union shops down south, when products became outdated. And when they weren't stuffing boxes with pink slips and eliminating good jobs—some of the best available to the denizens of small-town America—corporations seemed unable to pass up an opportunity to cut wages, gut benefits, or, the old standby, bust unions.

Each of corporate America's actions got a nod and an assist from the politicians. The body politic of the United States experienced a precursor to the Washington Consensus of the 1980s—the punitive monetary policy that the United States levied on developing countries confronting fiscal crises and in need of support from the developed world—when, beginning in the early 1970s, a cavalcade of GOP *and* Democratic administrations championed neoliberal economic policies. It wasn't so much an emerging political realignment, as Kevin Phillips argued in his prescient book *The Emerging Republican Majority* in 1969, but an emerging *economic* order.

One after another—Nixon, Ford, Carter, Reagan, Bush Sr., Clinton, Bush Jr.—U.S. presidents rammed through agendas of cutting taxes, deregulating industry, championing free trade deals, and supporting tight monetary policy. True, Nixon did make his infamous admission that he was a "Keynesian in economics now" in

1971, but in retrospect it was nothing more than a head fake—just enough to get him through an election.

I bring up George Clooney and Mark "Marky Mark" Wahlberg's box office disappointment *The Perfect Storm* because it is the most well-known movie on the topic of shipwrecks and adventure. (Sorry, *Pirates of the Caribbean* doesn't count, Mr. Depp.) I also use it because the title suggests that the film's subject was *the* voyage that defined the movie genre and to the exclusion of all others.

Wrong.

The Perfect Storm was about sex and machismo. The wreck of the *Edmund Fitzgerald* was a display of bravery, courage, and selflessness in which everyday, ordinary men, who took it upon themselves to ensure the Great Lakes economy remained strong and millions of hard-working families like their own, would also have good jobs and be able to support their families. In other words, the men of the Fitzgerald kept food on the tables and roofs over the heads of the American middle class, and they made sure every child had a shot at the American dream, whether they were a son or daughter of a shipping magnate, insurance company executive, small town lawyer, or waitress at Louis's Restaurant in Superior, Wisconsin.

The events leading up to and including the wreck of the *Edmund Fitzgerald* should be chronicled in history textbooks and included in the curriculum of every school in *every* state. This text will make this argument quite clear. *The Perfect Storm*, however, belongs in DVD format only and in the basement romper rooms of overstimulated teenagers.

I have an ensemble cast already in mind for the movie of this great American ore ship's story. It will be a blockbuster.

But first, the book.

It has been fifty years since the *Edmund Fitzgerald* sank into the icy depths of Lake Superior. It was more than just a shipwreck, more than the namesake of a staple ballad on classic rock radio stations, and even more than a painful memory that lingers among family and friends of the fallen. Incredibly, there are few books that truly tell a story worthy of the great iron ore ship's legend. Why? This question has overshadowed this project; I still don't have an answer.

What this book seeks to accomplish is to place the *Edmund Fitzgerald* in a bigger, more significant historical context than other works have done. The canon of Fitzgerald scholarship that has accumulated over the years has been essential to better understanding the mystery behind the sinking of the Fitz. This book is not intended to present an exhaustive review of the various theories behind the ship's demise. Instead, it is intended to apply the theories that have been developed and research collected to tell a more complete story of the Fitzgerald.

This book reflects not only research and scholarly works but also accounts of and lessons learned from important deeds and good works performed by family members, shipwreck enthusiasts, divers, and former sailors. Their work and stories have been critical in helping me tell a greater story of the ship.

Tim Nixon, former Great Lakes shipman (1976–87) turned attorney carefully catalogued contemporary news articles from 1975 to 1977 hoping this would help tell the Fitzgerald story. His work is published for the first time in chapter 4.

Attorneys Toby Marcovich of Superior, Wisconsin, and Steve Eckman of Independence, Minnesota, and their respective legal teams logged untold hours and road and flight miles to bring long deserved justice to the families. Their stories as well as revelations of another previously unreported Fitz conspiracy are included in chapter 5.

Tom Farnquist's monumental task of retrieving the bell from the ship's wreckage in 1995 is featured in chapter 9. He did more than raise the bell. He brought disparate parties and families together to honor the memories of their loved ones, bringing them much needed closure.

Evidence of a possible second body in addition to a first that has haunted the deep recesses of the internet and dominated discussion boards of shipwreck buffs at the wreck is also presented in this book. This is a sensitive topic because the wreckage is considered a graveyard and protected as such by Canadian law. Explorers are not allowed to dive the wreck, much less search through the wreckage for artifacts and evidence of human remains. Diving technology has improved remarkably since the wreck was last viewed thirty years ago, and it is particularly hard for explorers and shipwreck buffs to walk away from this artifact.

Naturally, this book begins on November 9, 1975, and chapter 1 presents a minute-by-minute account of the ship's last journey, based on the official U.S. Coast Guard report and eyewitness reports of the ship's departure and final minutes.

Chapter 2 lays the groundwork for making the connection between the sinking of the Fitzgerald and the congruent sinking of the national economy. It follows the narrative arc established in chapter 1 by recounting how the post–World War II economy led to the inequities of the present day. *Plus ça change*. The more things change, the more they stay the same.

Chapter 6 follows the arc of chapters 1 and 2, telling the story of the shipyard that berthed the Fitzgerald in 1958, as well as the rough-and-tumble, hard knock life of the Fitzgerald as she logged over one million nautical miles across the Great Lakes, hauling up to 27,402 long tons per trip for almost two decades.[6] The chapter chronicles the shoddy design and poor maintenance record of Oglebay Norton Corporation, the Fitz's owner, which were overlooked by governmental oversight agencies.

Over one billion years ago a rift within the earth's core exploded and forged the unique features of what would become Lake Superior, the second largest lake in the world. Untainted by invasive species and with a high calcium content, its pristine waters have preserved shipwrecks like the Fitzgerald for centuries, but the impending climate crisis will make the waters more dangerous for ships. Chapter 3 explains why the largely ignored Fitzgerald story is worth exploring and understanding.

Chapter 10 personalizes the story of the Fitzgerald by profiling four people from four different parts of the country, each of whom lost a family member aboard the ship. Each of their stories is unique, each story colored in its own way by the untimely passing of the person's loved one.

Chapter 4 picks up where chapter 1 left off. It is the evening of November 10, 1975, and a rescue effort for the Fitzgerald is underway. Just as Oglebay Norton Corporation failed its employees by allowing them to sail on an unseaworthy vessel and the American Shipping Bureau looked the other way when the Fitzgerald was literally falling apart, a botched rescue attempt fails the crew.

Chapter 5 features family members and legal cases to recover damages for the deaths of their loved ones. These dramas highlight the emotional ups and downs experienced by the Fitzgerald families, as their legal twists and turns prove a long and tedious road to justice.

U.S. shipbuilding capabilities should not be reserved for military purposes alone. Billions of dollars in federal defense outlays have no doubt strengthened the industry by employing and training thousands of shipbuilders from welders to pipe fitters to electricians. Chapter 8 suggests how a new public-private sector business model not unlike that of the past could rebuild the industry and spur a renaissance in American manufacturing.

In the leadup to the wreck of the Fitzgerald in 1975, the Democratic party was ascendent. But the Vietnam War (1964–73) and later the Watergate scandal (1972–74) stoked anti-Washington political sentiments in the body politic. Chapter 7 explains why states like Michigan—the heart of the Great Lakes economy—shifted decidedly

Republican in voting behavior. The failure to revitalize American industry was at the heart of this phenomenon.

Chapter 9 recounts the remarkable tale of how the bell of the Fitzgerald, or "the soul of the ship" as Deborah Champeau, daughter of third engineer Oliver "Buck" Champeau put it, was raised. This chapter explores a fascinating, heartwarming story that speaks to the spirit of the Fitzgerald. Not every story about the Fitzgerald is a tragic one.

The final chapter brings the story full circle. All is not lost. We can rebuild the shipping industry and, by extension, rejuvenate American manufacturing. Chapter 11 lays out a roadmap for how we can accomplish this worthy goal in a manner that honors the memory of the twenty-nine fallen crew members of the *Edmund Fitzgerald.*

CHAPTER 1

The Wreck of the *Edmund Fitzgerald*

IT WAS THOMAS WOLFE WHO TOLD US, "YOU CAN'T GO HOME AGAIN." AND SO IT WAS with the *Edmund Fitzgerald*. Only on rare occasions did this storied Great Lakes ore ship ever return home. Instead, the Mighty Fitz—as she was affectionately known—overnighted in Silver Bay, Minnesota, when she was upbound on the Great Lakes and in Toledo, Ohio, when she was downbound, resting along the docks of Lake Erie.

On the rare occasion she made it back home to Milwaukee, the *Edmund Fitzgerald* sat a few blocks from the headquarters of her owner, Northwestern Mutual. The insurance giant has stood tall along the skyline of Wisconsin's largest city just as it dominated competitors in the life insurance market for years. So it's not surprising that when Northwestern Mutual paid the Great Lakes Engineering Works of Detroit to build the Fitzgerald, they had just one requirement: make her the biggest ship to ever sail the Lakes.

The ship's namesake was the CEO of the company. Edmund Fitzgerald did not come from modest means. His father was a renowned shipbuilder; his maternal grandfather was a respected manufacturer and industrial leader in Milwaukee who served in the Dwight D. Eisenhower administration; and his paternal grandfather was a ship captain.[1] Nevertheless, unlike today's scions of rich parents, Fitzgerald's father would make sure his son earned his keep. Thus began a career-long climb to the C-suite of one of the country's largest insurance companies and one of the Dairy State's biggest employers.

Despite getting out of the family business, Fitzgerald never lost the sailing bug. And he wanted to do right by the family, not just carrying on the family name but immortalizing their name in a great iron ore ship. Fitzerald wanted to make sure it was not only the largest ship but also a well-built and safe ship for its crew. Alas, you cannot have small budgets that cut corners and a high-quality product; it must be one or the other. It was tradition for ships to be named after the owning company's CEO or president. The *Arthur M. Anderson* was named after the CEO of U.S. Steel. The *William Clay Ford,* built six years before the *Edmund Fitzgerald* in 1952, was named after Henry Ford's grandson and member of the Ford Motor Company's board of directors. But despite all of these male namesakes, ships like the Fitzgerald are referred to as "she" or "her," as sailors have been wont to do over the course of history.

After loading up taconite ore in Silver Bay, the Queen of the Lakes, as the Fitzgerald was also referred to, would travel across Superior, through Huron and into Erie, where her cargo would be unloaded in Toledo, Ohio. From there, the ore would be hauled by rail car to steel plants in Ohio and Kentucky. Then, she would return to Silver Bay for another run. And another. And another. She repeated the journey seven or even eight months out of the year. The Fitzgerald's role was the third stage of the steel-making process. Pure iron ore had long since been extracted from the open pit mines of northern Minnesota and Ontario, Canada, in the 1950s, shortly after the ship- and plane-building demands of World War II had drained the mines dry. Companies were now mining taconite, which contained only 25 to 30 percent pure iron. Nevertheless, it would still get the job done by feeding the hungry forges of Detroit, Cleveland, Pittsburgh, and throughout Kentucky.[2]

On a beautiful June morning in Superior, Wisconsin, in 2023, almost forty-eight years since the *Edmund Fitzgerald* sank, Bob Hom, a former sailor of the Great Lakes, offered me a chair on his freshly laid asphalt driveway. The sun was brilliant, and the black driveway radiated its warmth. He could remember vividly the day and moment the Fitzgerald left the port of Superior for the last time. Hom would have been one of the last people on land who saw the ship before it sank.

"My wife and I drove over the bridge to go to Fairlawn Museum in Superior and we could see the Fitzgerald leaving out of the Superior port. It was a beautiful day like this. A gorgeous day," Hom told me.[3] On November 9, 1975, Bob and his wife, Peg, were going to meet up with Fitzgerald Wheelsman Eugene "Red" O'Brien, a good friend of the couple. Bob had sailed with Red as well as other Fitzgerald crew members on other boats. Red would have to take a rain check on dinner because the Fitzgerald was departing early. A storm had been building over the Oklahoma panhandle the day before. The system was estimated to hit just south of Lake Superior by the early evening of the following day.[4] Best to get ahead of the storm.

Hom recalled his phone call with Red like it was yesterday: "[Red] called and I said, 'Where are you?' 'Well, I'm in Superior,' he said. 'So, I'll come see you,' I said. 'No, we're just about ready to leave. This is our last trip and when I get done, I'm going to go up to Minneapolis and spend a week with my sister. And then I'm going to come and spend a week with you and Peg in Duluth.' 'Okay, I'll see you in two or three weeks.' And that was it."

Bob hasn't had much chance to talk about the Fitzgerald. It was a long time ago and "most of [his] friends and sailors are gone." "I still get a little misty eyed. I'm sorry. It's only been what, fifty years," Hom deadpanned as a tear or two welled up in his eyelids. "I don't talk about it very often. I don't know anybody else you could talk about it with." Peg, tending to flowers along the house, looks back to check on her husband to make sure he's okay. She knows how he feels. Red was her friend, too.

While a severe storm was gathering in the southwest on November 9, 1975, it was certainly not unusual for that time of the year. The *Edmund Fitzgerald* had handled plenty of storm systems from the south where the winds and warm fronts made weather more hazardous. Not as bad as a nor'easter, but trouble enough.

Fitzgerald captain, Ernest M. McSorley had a well-established reputation as a rough-weather captain who never blinked at going into a punishing storm. He had a job to do, and he took it quite seriously. "You don't make money sitting in port," McSorley, the quintessential company man, would say.[5] Decades at the helm of the Fitzgerald and nine other ships showed on his face.[6] Weather-beaten and serious, a smile that betrayed a total unwillingness of suffering fools. Certain legions of sailors and first mates learned that lesson the hard way.

Third Mate Michael Armagost was accustomed to shipping out of Silver Bay, Minnesota, which was a about a two-hour drive from his home in Iron River, Wisconsin. By comparison, Iron River felt like an outer-ring suburb of Superior. Armagost felt bad leaving First Mate Jack McCarthy at the dock to handle the loading, which would take most of the morning, but family was family. He had a wife, Janice, and two young children, Michele and Christopher, and valued his time with them.

After dropping Mike off at the dock, Janice took the kids for burgers. While finishing up their late lunch, they watched the Fitzgerald sail off in the distance. The three of them waved goodbye to their dad and husband, sad to see him go but anxious for the coming off-season when they could celebrate Christopher's third birthday and enjoy Thanksgiving dinner together.[7]

Armagost felt bad leaving McCarthy in part because Captain McSorley didn't like to spend any more money than he had to on these trips, and he leaned heavily on his fellow sailors to prep the ship for them. Having his mates aboard to ready the ship for sail meant he didn't have to pay deckhands double time because it was a Sunday.[8]

McSorley ran a tight ship, and his wife enjoyed the bonuses for making more runs with heavier payloads beyond his quota. McSorley was good at running the ship, and his wife Nellie proved it. "I always remember my aunt Nellie walking in with her big fur coat [at Thanksgiving or Christmas]," McSorley's niece Kay McSorley recalled to me. "Just glamorous . . . I always was amazed by the fur coat."[9]

Watchman Ransom "Ray" Cundy of Superior may have been slipping out of the local bar. "Ransom tended to drink his paychecks," Superior attorney Toby Marcovich told me, although he never failed to pay his lawyers' fees.[10] Toby used Cundy's fee payments as a mnemonic device to forever remember his name. "Ransom, as in money," he told me. It's one reason the seasoned small-town jack-of-all-trades (and master of all of them) lawyer would accept Doreen Cundy's case against Oglebay Norton Corporation, the shipping company that ran the *Edmund Fitzgerald*. She would be lucky to work with him; Toby had won million-dollar settlements, gotten someone off a murder rap, and labored through the occasional probate estate cases.

The description of Cundy was not necessarily inaccurate, but it wasn't exactly fair either. The year before in 1974, Cundy's daughter was shot to death by her husband in a horrific murder-suicide.[11] He was once the life of the party—or in this the case the ship—but the tragedy weighed heavy on him. He never regained the spark in his step nor his wry, trademark sense of humor. Nonetheless, his daughter, Cheryl Rozman, remembered him fondly as a devoted father with a "tender heart" and one who worked hard, made friends quickly, and always kept those friends close.[12]

But because of his drinking, Cundy's wife couldn't rely on him for a steady income. She took up work waiting tables at Louis's Diner, which Toby frequented. Louis's was the beating heart of downtown Superior, where people of all walks of life gathered—sailors, lawyers, doctors, farmers, longshoremen. This was where Toby met Doreen. The relationship ensured Doreen had a good lawyer on hand, in case she ever needed one.

As the *Edmund Fitzgerald* steamed out of the port on November 9, 1975, Bob and Peg Hom, just like Janice and her children, waved farewell from far off in the distance. The Fitzgerald's final voyage had begun.

Armagost, Cundy, McSorley and the rest of the crew were hoping this would be the last trip of the season. But if the weather was any indication, they could be going into December, because it felt like an Indian summer. One of the boat loaders, Clarence Dennis, would have agreed with Bob Hom about the day. "It was a nice day. It could have been a little cloudy, but it was not a stormy day," Dennis said.[13] Winds were picking up and there could be some storms in the offing, but that was nothing unusual. "We were indicating a general increase in winds and an increase in the waves," according to Raymond Waldman of the National Weather Service. Forecasts would change in the coming hours, but for now, all was quiet.[14]

Although in 1975 meteorological equipment and techniques were not archaic, ore freighters like the Fitzgerald were woefully underequipped, and there were no dedicated fixed entities along the way to collect data and make weather observations. Crews relied on ships ahead of the weather to gather data and report back. It was a less-than-ideal system, and the National Weather Service would agree. "We have very few places from which we get weather information from a fixed point on a continuous basis, so we have an urgent need for that kind of information," Raymond Waldman of the National Weather Service in Chicago said.[15] The truth was that the union had fought tooth and nail for these and other workplace safety improvements. Employers were either too cheap or too oblivious to find ways to improve travel efficiency, protect workers, safeguard expensive cargo, and save money—or all four.[16]

Workplace safety had not been taken as seriously then as it would be later. The Occupational Safety and Health Act (OSHA) had recently been adopted (1970). Automation and mechanization not only led to job loss but also sped up production processes that "increased dangers" for workers toiling alongside assembly lines. The proof was in the pudding: occupational injury incident rates were a staggering threefold greater in 1975 (9.1) than in 2018 (2.8).[17]

Charts proved useless, but not because they weren't good sources of information. Ship captains were not unlike dads who refuse to use road maps to make their way around the country on family vacations. "I had captains yell at me for using radar, search lights, and looking at charts," former Great Lakes sailor (1976–87) and Green Bay, Wisconsin, attorney Tim Nixon told me; although in this instance it would not matter because the chart available on the *Edmund Fitzgerald* was fifty-three years old.[18]

The reason for the captains' unwillingness? "According to the captains, you didn't need it. Didn't like you talking on the radio; didn't like you telling anybody anything," Nixon said. Rand McNally doesn't tell dad how to drive the family minivan, either.

The *Edmund Fitzgerald* was heading to Detroit rather than her usual lower lake destination of Toledo, Ohio, because that day she was sailing for National Steel and not Oglebay Norton Corporation. Ship companies had mutual aid agreements. If one needed a load of taconite at a particular time but did not have the capacity to fill the order, they could tap the resources of another company. It wasn't too common, but it did happen. In 1975, the Fitz loaded up at the Superior ore docks on four instances, or approximately 10 percent of the time.[19] The Fitzgerald made on average forty-four round trips each season.[20] In this instance, National Steel needed a load of taconite for its steel mill in Detroit but didn't have a ship in the upper lake.

Oglebay Norton Corporation got them out of their pinch. Yes, they were competitors, but they were in the same league, if not on the same team. Foreign competition was crushing the steel industry, and it was gobbling up more and more of the U.S. domestic car market. Collectively, steel companies had an incentive to cut costs as much as possible to remain competitive with the steel giants overseas such as Japan and South Korea. From 1950 to 1976, U.S. steel production fell by over one-half—from 50 percent to 20 percent of world production. By the 1970s the "only positive export trend . . . material dominance which underlay international economic strategy vanished." This is to say nothing of a comprehensive, economic-wide industrial strategy. Worse, when elected officials had a good idea or two to shore up industry and jobs, the country's editorial pages and elite were assailing "protectionism" though "ignoring the foreign subsidies or government ownership" that created industry's problem in the first place. Trade policies were also "dominated by strategic political considerations" as opposed to serious domestic, economic concerns typically in an effort to bolster counter weights to communist spheres of influence in southeast Asia.[21]

The winds picked up as the day wore on. Cedric Woodward, pilot of the Swedish-flagged *Avafors*, and Captain Jesse Cooper of the *Arthur M. Anderson* were both on the lake that evening, though they were coming from opposite ends of the lake. Each had a different take on the weather. "I would consider it a severe local storm because it was intense, but it was not a vast wide two-day storm," Cooper said.[22] "It was very severe . . . it was one of the biggest and wildest seas I have been in," Woodward said.[23]

The difference in perspective can be easily explained. According to Ledolf Baer, director of the National Oceanographic Services office, "There is a great deal of variance between eyeball estimates."

"What studies show is if you are on a big boat, you tend to underestimate the waves. . . . [The studies] showed that what people think they see from a ship, what they talk about wave height, is the significant height," said Baer.[24]

Woodward's *Avafors* was a saltwater ship and built for the rolling seas of an ocean, not the mercurial, pounding waves of Superior. And Woodward had the added perspective of sailing oceans and not just lakes. The waves appeared larger and felt worse to Woodward than what Cooper saw and experienced. The inexact science of "eyeballing" wave size proved even more dangerous because that was all captains had to go on. Ships on the Great Lakes did not have any instruments to measure the size of waves.

Just nineteen minutes after the Fitzgerald left safe harbor, the National Weather Service issued gale warnings. They were upgrades from the report they received at 2:20 p.m. but nothing of great concern; the reports were always subject to change, sometimes at a moment's notice. In the Midwest, there is a saying, "If you don't like the weather, wait an hour."

At 4:15 p.m., McSorley spotted the *Arthur M. Anderson*, the other ore freighter, leaving her port of Two Harbors, Minnesota, about half an hour south of the Fitzgerald's normal upper lake port of Silver Bay. The Anderson was about fifteen miles behind the Fitzgerald. Just to be cautious, McSorley and the Anderson's skipper, Jesse "Bernie" Cooper, made it a point to stay in contact. This was particularly important for the Fitzgerald because they were heading north and east, away from the storm. The Anderson would get the weather first and would be the Fitzgerald's main source of weather information from here on out—more so than the National Weather Service.

The two ships crossed paths around 7:00 p.m. The Anderson would be sailing to the south of the Fitzgerald, while the latter would be taking a more northerly route. They would remain in this formation until the following afternoon at 12:30 p.m. and then switch once more around 3:30 p.m., when each was approaching Whitefish Bay on the far eastern edge of the lake and thereafter the Soo Locks at Sault (pronounced "Soo") Ste Marie, which connected Superior with Huron via the Saint Mary's River.

The next important milestone for the Fitzgerald was passing Isle Royale in the lake's northern waters around 1:00 a.m. This island was a leftover chunk of earth rock on the lake's basin that the glaciers missed when retreating north some forty thousand years earlier. It was of good size and hard to miss.

While the weather conditions grew in severity, the Fitzgerald made her regular weather report at 7:00 a.m., but this would be her last one. She would miss the regularly scheduled 1:00 p.m., 7:00 p.m., 1:00 a.m., and 7:00 a.m. reports. The Fitzgerald reported that wind speeds were at thirty-five knots and waves were as high as ten feet. She also reported that their arrival at the Soo Locks would be "indefinite, due to the weather."[25] She would augment her route by tacking north to be closer to land, seeking cover from stormy waters. Still, there was nothing to suggest things were seriously awry. Captains and crew were well accustomed to the weather this time of the year; November was the worst month for sailing.

On Monday, November 10, at 1:00 p.m., the Fitzgerald was within eleven miles of Michipicoten Island, located in the north-central portion of the lake. This was as close to the shore as the Fitzgerald could be and still make the Soo Locks without getting too far off its direct route. McSorley had to decide: safety or efficiency. Ships carried only enough fuel to make port and had to conserve resources. There was also a tradeoff: the more fuel a ship carried, the less cargo it could haul. Battling rough seas burned up additional precious fuel. Timing was also important to captains because saving time also meant bonuses, though crew members had no such incentive.

At 2:45 p.m., the Anderson changed course to avoid the Six Fathom Shoal, which was located just north of Caribou Island and ninety miles away from Sault Ste Marie. Cooper noted that the Fitzgerald held her position and "passed close" to the shoal. According to an account taken during a U.S. Coast Guard investigation the following year, Cooper told his first mate that "the Fitzgerald was closer to [the] shoal than he wanted the Anderson to be."[26]

And just as older captains refuse to consult charts, they also avoid telling each other what they should or should not do with their own ships. Cooper's testimony does not say if he warned the Fitzgerald that she appeared to be getting too close to the shoal, nor does it indicate whether the question was asked at all. According to Jim Woodward, son of *Avafors* captain Cedric Woodward, Cooper refused to allow his first mate to call the Fitzgerald and warn McSorley he was heading for the Six Fathom Shoal.[27] "My dad was on the *Avafors* that night and they were talking of the event of that night and that's when the mate told my dad about it." The reason? "Another captain doesn't sail another captain's ship," Woodward recalled.

The Fitzgerald did not avoid the shoal. For many historians and Fitzgerald buffs, this was the decision that triggered the chain of events leading to the ship's demise. The decision would be central to one of the longest sustaining and credible theories, one that Cooper believed until the day he died; the ship bottomed out on the Six Fathom Shoal and tore a hole in the hull. "I don't care what anybody says . . . she had either bottomed out or had a stress fracture in the hull," Cooper stated in the government's official after-action investigatory report, referring to the notorious Six Fathom Shoal near Caribou Island that had grounded other boats in the past.[28] "She was sinking from that time on."

Ric Mixter, a documentarian, former TV reporter, radio DJ and Fitzgerald enthusiast who has studied the wreckage since he made a dive to the Fitzgerald in 1994, would not agree. Mixter would argue that there was no sign of underlying damage to the stern despite the fact that—confirmed by his first-hand observation—it rested in an inverted position on the bottom of the lake. Mixter also argues that there were no signs of the ship running aground at Six Fathom Shoal,[29] although this is a rather dubious conclusion since the shoal is one mile long and was never fully inspected. Moreover, shoals—like any matter on a lake bottom—are susceptible to erosion and any evidence of a grounding would have disappeared long ago.

Meanwhile, a low-pressure system from Canada was washing over the lake and mixing with water still warm from the summer, producing a massive snowfall. Combined with high winds, sailing conditions quickly deteriorated. That was enough for the Coast Guard to direct all ships headed to the Soo to anchor near shore. The locks were closing, and the weather was getting worse. But the Anderson and Fitzgerald soldiered on.

Forty-five minutes later, at 3:30 p.m., McSorley reported top-side damage. This was the first report from McSorley that the ship was sustaining damage. According to McSorley, the deck's rail fence broke, vents on the deck were damaged, and the boat was taking on a list.

> **MCSORLEY:** Anderson, this is the Fitzgerald. I have sustained some topside damage. I have a fence rail laid down, two vents lost or damaged, and a list. I'm checking down [slowing speed]. Will you stay by me til I get to Whitefish?
>
> **COOPER:** Charlie on that Fitzgerald. Do you have your pumps going?
>
> **MCSORLEY:** Yes, both of them.[30]

McSorley had had enough. He was doing the sailing equivalent of stopping at the nearest gas station to ask for directions.

The three damage reports were significant because they were highly unusual. It was almost without precedent for them to occur at the same time.

First, fence rails on the side of an ore ship rarely come down. They are built to last and withstand the worst possible weather conditions. The rails consist of three 3/8-inch wire cords strung through three-inch-thick stanchions thirty inches tall. Sometimes one cord will snap. Once in a blue moon the second strand will break. But it is rare that all three strands break and the rail collapses. If it were to snap off completely it is likely because of hogging, according to Cooper. Hogging typically occurs when cargo is not evenly distributed. On the Fitzgerald and other Great Lakes freighters, cargo holds are separated by walls, but the dividers are not indestructible—far from it. In rough seas, shifting cargo can break through the walls and spill into other holds.

Second, the damaged vents—which had likely popped off, according to testimony by Cooper—suggested that intense water and air pressure from below popped the vent like an engine overheating and blowing a gasket. The engineers would quickly turn on the ship's two ballast pumps. Ballast pumps pull water out of the ballast tanks when the ship comes to port. Water pours out of a hole on the side of the ship as if it were relieving itself. They can also be used to pump out water coming into the ballast tanks while sailing. The cargo hold is another matter. On a full load, and especially in heavy storms, pumps in the cargo hold are next to useless. Pumps get clogged up and water that does get into the hold usually stays there because taconite, which consists of permeable sand silica, acts like millions of little sponges absorbing water. It weighs the ship down considerably.

Third, McSorley disclosed "a list," which occurs when damage is done to one side of the ship, including a tear in the hull. That is, the ship begins to drift just slightly in a certain direction without the control of the helm. Cooper would soon after describe the list as a "starboard list"—the side of the ship facing Six Fathom Shoal. That would mean it would be quite possible that it had run aground and sprung a leak. For some, including Ric Mixter, the details are neither significant nor accurate. Mixter likens Cooper's claim of a starboard list to the telephone game, as telling and retelling a story allows it to take on new details, shapes, and sizes.[31] Another Fitzgerald enthusiast, Jeff Thomas, notes the discrepancy of Cooper referring to a starboard list after the fact but gives it short shrift.[32]

While the argument that Cooper misrepresented the occurrence certainly has some merit, Cooper's detail should not be discounted. The events of November 10 were more significant and lasting than anecdotes and jokes told at a cocktail party. In the eighteen years since the ship's sinking, Cooper had more than enough time to turn over in his head the events of November 10. He made this statement for the Coast Guard report and maintained his conclusion until the day he died: "I don't care what anybody says. At 3:10 in the afternoon, she had either bottomed out or had a stress fracture in the hull. That's the only two possibilities. She was sinking from that time on."[33]

In *The Trial of the Edmund Fitzgerald*, editor Michael Schumacher cited a transcript from the U.S. Coast Guard investigation's interview with Cooper, which had heretofore not been published. According to Schumacher, Cooper explains why he considered it a starboard list: "I was up in the wheelhouse when [McSorley] called, and he told me that his fence rail was laid down and that he had snapped off two vents and he was taking water through those vents into the tanks. He didn't say starboard or port list, but I would assume it was a starboard list, because the sea was on our starboard side a little bit."[34] If he was not certain it was a starboard list, why didn't he change his mind in the eighteen years between the U.S. Coast Guard interview and Mixter's interview in 1993?

It is the job of a captain to gather information and make decisions based on real-time information. Cooper had been at the helm of a Great Lakes ship for ten years and had been sailing for nearly thirty-five years.[35] He, more than anyone else—and that includes amateur historians—would be in the position to know what was going on during the afternoon and evening of November 10, 1975. And his recall was based on recorded information—a complete radio transmission and transcript of the exchange—and not on mere memory. Indeed, the transcript is printed on a wall-size painting at the Valley Camp Museum in Sault Ste Marie for thousands of visitors across the country and world to read every year. In addition, there have been scores of books, pamphlets, scholarship, and tourist trinkets that have been written and made over the years. Now there is an entire cache of internet searches of the transmission and transcript. Cooper could easily have consulted these, reflected on the moment, and come to a different conclusion. But he didn't. Cooper died in 1993.

At 4:10 p.m., the Fitzgerald reported that its radar was not working. The radar was the most reliable equipment the Fitzgerald had for navigation. Incredibly, both of the ship's radars were not working even though they were on different circuits for the purposes of redundancy. The Fitzgerald would now have to rely entirely on the Anderson for radar positioning.

It was also around this time that McSorley asked for "navigational assistance"; in other words, conditions were not improving. Anderson's first mate, Morgan Clark, was back on watch (the second mate had assumed watch after noon) and quickly obliged. But by then the Fitzgerald had already come within striking distance of Six Fathom Shoal. For McSorley, it was too little, too late. For Clark, it was "told you so."

At 7:10 p.m., the Anderson made final contact with the Fitzgerald. The last words exchanged by First Mate Clark and McSorley would live in infamy in the annals of American shipwreck folklore. Not even the most infamous of shipwrecks, the *Titanic,* had a more memorable sequence at the end of its life.

ANDERSON: Fitzgerald, this is the Anderson. Have you checked down?

FITZGERALD: Yes we have.

ANDERSON: Fitzgerald, we are about 10 miles behind you, and gaining about 1 1/2 miles per hour. Fitzgerald, there is a target 19 miles ahead of us. So the target would be 9 miles on ahead of you.

FITZGERALD: Well, am I going to clear?

ANDERSON: Yes. He is going to pass to the west of you.

FITZGERALD: Well, fine.

ANDERSON: By the way, Fitzgerald, how are you making out with your problem?

FITZGERALD: We are holding our own.

ANDERSON: Okay, fine. I'll be talking to you later.[36]

The transcript suggests that McSorley was totally oblivious of the ship's condition, although in McSorley's defense, he wouldn't have been able to tell if the waves were getting larger or he was just getting lower into the water. Said Cooper: "He gave no indication that he was worried or that he had a problem or there was something he couldn't cope with. There was no excitement whatsoever. This was a problem, but it was under control. This is what you would assume from the way he talked, that there was no problem."[37]

Roy Anderson, second mate of the Anderson concurred: "I was told that he was not in any standing danger. He was not having any problems, and I believe if he was, that he was to call me on that."[38] In the Coast Guard report, Cooper says, "I firmly believe that he had a damaged ship and didn't know how damaged she was." When McSorley chose not to go "right up on the beach in Caribou" it told Cooper that he was unaware of the ship's true condition. And Cooper was positive about that conclusion. Rare as it may be, "he would have put [the boat] on the beach. I am sure of it," Cooper said.[39]

It is not clear if Cooper's assessment was made retrospectively or if he believed it at the time. If the latter, it begs the question, Why didn't Cooper warn McSorley? Was it because "a captain doesn't tell another how to sail their ship?"

First Mate Morgan Clark could just make out what he believed were the lights of the Fitzgerald. Flickering. On and off. On and off. On and off. And then off.

Clark took a breath, rubbed his eyes, and strained to find the lights again. It was like he was trying to will them back on. He could now see the lights of the *Avafors*. The Fitzgerald had been in between the Anderson and the *Avafors*. If he could see the *Avafors*' lights, he should have been able to see the Fitzgerald. But he didn't.

She was gone.

CHAPTER 2

The Wreck of the American Economy

JUST HOW THE CREW OF THE *EDMUND FITZGERALD* FOUND THEMSELVES ON A SHIP in the condition that it was in and on a lake in dangerous weather conditions takes some explaining. Primarily, it was an end result of corporate strategies as economic conditions grew worse in the latter half of the twentieth century. Inflation was rearing its ugly head, foreign competition was crushing domestic producers, and American industry was woefully underinvested in research and development and unresponsive to consumers and unprepared for the repercussions of those inactions.

Not surprisingly, companies—including Oglebay Norton Corporation, the parent company that operated the Fitzgerald—were taking extra chances and leaning on captains to disregard safety issues to meet bottom lines, like sailing when ships were not in good shape or weather demanded captains make landfall and not press on. But the owner, Northwestern Mutual, proffered a triple net lease to indemnify them of any liability should something happen to the crew or the ship. Since the business model of insurance companies like Northwestern Mutual is based on denying claims, the contract was airtight.

Sure, pressure was always on workers to meet if not exceed quarterly net income goals, but entire industries were on the brink of extinction and the first round of plant closures and layoffs pummeling company towns were underway, especially in the Upper Midwest.

Too many stories are punctuated with the epitaph "It could have been avoided altogether"—like faulty foreign policies that lead countries to war and young men and women to the front lines of battle or cheap bosses who cut corners rather than invest in their products and their employees and hand out pink slips as stocking stuffers at the end of the year. The *Edmund Fitzgerald* was no exception.

The story of the wreck of the Fitzgerald begins not on the fateful afternoon of November 9, 1975, in Superior, Wisconsin, but as soon as the guns fell silent in Asia and the South Pacific after World War II. Back then not everyone was out for themself. There was more than profits and stock dividends; there was community and a way of life that fostered an inclusive society where the American dream was within reach for the average American family. In short, the economy worked for all: CEOs, plant workers, captains, and deck hands alike. Many of the "economic royalists" that FDR had called out in the war years had been reined in. Income inequality was at an all-time low.[1] The country was not without political and economic leadership; folks who heeded the warnings of unbridled capitalism, the "animal spirts" in a market economy, as John Maynard Keynes famously articulated on the eve of the Second World War in 1936. One man who acknowledged Keynes's forecast was a contemporary who respected him and relied on him often for advice to do his job.

No sooner had President Franklin D. Roosevelt won World War II than the American economy reached its zenith. One in three workers in the private sector were in a union. Mean household income was at $25,000, a twofold increase from the start of his administration in 1933. And unemployment was at a peacetime low of 3 percent. Had FDR lived beyond 1945, these numbers could very well have grown more. The late president had envisioned "a sweeping post-war expansion of the New Deal," according to labor historian Jack Metzger.[2]

A decade earlier, in 1933, Roosevelt had stared into the abyss of American capitalism. One in four Americans were out of work, real GDP bottomed out at $817 billion, and the Dow Jones industrial average sat at 1,341. The New York Stock Exchange indices had been at 4,200 at the start of the Depression in 1929.[3]

In the 1930s the Democratic Party was threatened by the far left and radical undemocratic forces including the Communist Party USA. From within FDR's party,

Louisiana's populist U.S. Senator Huey Long offered a competing, socialist-oriented program, entitled Share Our Wealth.[4] The massive wealth redistribution proposal capped personal wealth and limited annual income. The excess would be redistributed as a universal basic income program, housing subsidy, free college education, and old-age pensions. Roosevelt feared it would have untold negative repercussions for a market-oriented economy. It went way too far. FDR was a politician through and through and notorious for avoiding key decisions, most famously the selection of vice president at the 1944 convention. Nevertheless, the New Deal was the genuine article, and it was Roosevelt's creation. Its legacy was a successful response to what early historians categorized as an "economic catastrophe and political crisis."[5] And FDR wasn't just a winning politician that partisans rallied around. Historian Nelson Lichtenstein refers to the president's supporters as the "Roosevelt cult."[6]

Roosevelt refused to capitulate to the far-left forces and move the United States off its moorings as a capitalist economy based on innovation, industrialism, and entrepreneurship. While failed attempts by Progressives in the early part of the century to establish a minimum wage, repudiate the gold standard, achieve workplace safety measures, and establish an enduring labor movement made clear the system's many faults, the country's economic system had proven to be the winning formula for conquering the world economy. Endowed with great material wealth and breathtaking ambition personified by the likes of Andrew Carnegie, John D. Rockefeller and Cornelius Vanderbilt, the United States overtook China in 1890 and claimed the mantle of world's largest economy.[7]

Still, there had to be a better way, an economic system that would create a sustainable order that would comport with the democratic ideals established by the country's Founding Fathers: life, liberty, and the pursuit of happiness.

FDR's sweeping electoral victories in 1932 and 1936 gave him a political mandate necessary to implement a far-reaching economic agenda that would remake the American economy. It would stop the bank runs that had crippled American finance. It would put the unemployed back to work through the Civilian Conservation Corps, Works Progress Administration, and other state-sponsored job-generators. It would bring business and labor to the table under the National Industry Recovery Act of 1933 to hash out pricing for basic commodities and products that would ensure fair prices for employers and higher wages for workers. Roosevelt's inner circle of economic advisors put stock in the "centrality of investment" and "value[d] predictability, stability and order." While FDR never endorsed any particular "concept of political economy" on the campaign trail in 1932, he had "a broad vision of economic rationalization that business planners and state planners seemed to share."[8] The beginnings of managerial economics were in place.

FDR's fellow New Yorker and U.S. senator Robert Wagner championed legislation that would codify labor's right to remain at the bargaining table with employers through the National Labor Relations Act of 1935 (a.k.a. Wagner Act). Employers would recognize the right to organize unions and collectively bargain on topics such as wages and benefits and workplace hours and conditions. Its significance could not be overstated, "it established the most democratic procedure in U.S. labor history," according to labor historian James Gross. "Collective bargaining," said Wagner, "is at the heart of the struggle for the preservation of political as well as economic democracy."[9]

To be sure, New Deal legislation put the American economy back on track. But only when FDR assured businesspeople and the investor class that the New Deal was "a response to a national emergency" in the summer of 1935 would the markets respond favorably, placing the economy on terra firma going into 1936 and FDR in the White House for another four years. Roosevelt said (in 1935), "The basic program [New Deal legislation] however, has now reached substantial completion and the 'breathing spell' . . . is here—very decidedly so."[10]

World War II began in Europe in September of 1939 when Nazi Germany invaded Poland, and France and Great Britain declared war on the aggressor nation. The United States became involved little by little, first by supplying armaments through the Lend-Lease Act, by which America lent military equipment to Great Britian from artillery to ships and planes, and later with boots on the ground. As the United States became the free world's arsenal of democracy, the American economy recovered from a double-dip recession in 1937 and 1938, and FDR breathed a not-so-subtle sigh of relief. Combined with his New Deal program, war expenditures gave the U.S. economy the necessary economic stimulus to fully recover from the Depression. FDR's political and economic legacy was intact.

With the end of World War II in sight, FDR delivered what would be known as his Second Bill of Rights speech during his State of the Union address of 1944. The country was consumed by the war effort and over eleven million Americans mobilized and fighting abroad; FDR returned to a campaign pledge made twelve years earlier.[11] At a visit to the Commonwealth Club of San Francisco in fall of 1932, Roosevelt had told its attendees, "The task of government in its relation to business is to assist the development of an economic declaration of rights, an economic constitutional order. . . . It is the minimum requirement of a more permanently safe order of things."[12]

On a balmy evening on January 11, 1944, as shallow patches of fog lurched through the streets of the nation's capital, FDR began to lay the groundwork and fashion the framework for an economic bill of rights for the American people. Roosevelt was beginning his twelfth year in the White House and his body was tiring. His declining health kept him from making the trek up Pennsylvania Avenue to the Senate chambers, and he delivered this address as a fireside chat from his White House study.

American soldiers had been slogging it out over desert landscapes of North Africa, through treacherous sea lanes of the South Pacific, and in the skies over western Europe. That Roosevelt could shift focus to the peacetime equivalent of war spoke to the gravity and seriousness of the matter at hand.

FDR's aim was to win the peace. War would rage for another nineteen months before the guns would fall silent. But this wartime president who had led the country out of the Great Depression knew planning for a strong robust economy for all—not a few—would need to begin in earnest. FDR argued that citizenship included not just political rights but substantive economic rights: the right to a job, to earn enough to make a living, to have a decent home, to have health care, to not live in economic insecurity, and to have an education. This would transform the premises of our nation's founding document and thus transform the rights to which Americans were entitled.

> This Republic had its beginning, and grew to its present strength, under the protection of certain inalienable political rights—among them the right of free speech, free press, free worship, trial by jury, freedom from unreasonable searches and seizures. They were our rights to life and liberty.
>
> As our nation has grown in size and stature, however—as our industrial economy expanded—these political rights proved inadequate to assure us equality in the pursuit of happiness. We have come to a clear realization of the fact that true individual freedom cannot exist without economic security and independence. "Necessitous men are not free men." People who are hungry and out of a job are the stuff of which dictatorships are made.[13]

Those properly articulated words, uttered with his commanding patrician accent, echoed beyond his study and into the past. They sounded like Lincoln's Second Inaugural Address, which had been delivered on the other end of Pennsylvania Avenue eighty years earlier. At that time, Lincoln had beseeched his countrymen not to rejoice in victory but to prepare for the long, hard, cold reality of reconstructing a completely destroyed nation and establishing a lasting peace between north and south, black and white—a peace that continues to elude the Union even today. "Let us strive on to finish the work we are in to bind up the nation's wounds, to care for him who shall have borne the battle and for his widow and his orphan, to do all which may achieve and cherish a just and lasting peace among ourselves and with all nations."[14]

FDR had been trading telegrams with British Prime Minister Winston Churchill and Soviet Union General Secretary Joseph Stalin, planning summits to carve up the globe, distribute the spoils of war, and ensure a lasting peace. FDR wished to be the one *true* Wilsonian president in American history, after President Wilson failed to win the peace in the aftermath of World War I. The League of Nations, the supranational organization that was his brainchild, fell apart, while the war reparations imposed on Germany backfired and that nation regained its strong economic and military footing under the Nazi dictatorship. It only made sense for Roosevelt to plan for his country's economic future—a democratic future—as well.

Of course, Roosevelt would need the consent of the U.S. Congress. He had jousted with senators and representatives to profound effect and was as much a master legislator as a successful executive.

But Roosevelt would not see the end of the war nor the implementation of the final phase of his economic agenda. This would be left to his most recent vice president, the untested though underestimated Harry Truman. The Missourian haberdasher cum politician would carry the torch.

Unfortunately, the union movement would begin its decades-long descent when Truman assumed office in 1945. Indeed, Truman vetoed—and the U.S. Congress overrode on a bipartisan basis—the 1947 Taft-Hartley Act, an anti-union law that decimated union rolls and whose lingering legacy, the so-called right-to-work provision, roils state legislatures to this day. Such laws are typically the first order of business of legislatures and governorships when they flip to Republican control.

Truman doubled down on his working-class street cred and nationalized the steel industry in 1952 when companies refused to grant wage increases to striking workers.[15] However, the companies fought back, and ultimately the U.S. Supreme Court sided with industry 6–3, ruling that Truman lacked the constitutional authority to seize private industry. All concurring justices were FDR or Truman appointees.[16]

In 1945 Democratic Texas Congressman Wright Patman introduced the Full Employment Bill. Congressman Patman was a curious combination of racist politics and economic progressivism. He was an original signatory to the *Southern Manifesto*, a document authored by southern legislators in response to the *Brown v. Board of Education* ruling in 1954 that declared segregation unconstitutional. It promised to use "all lawful means" to reverse the decision. But Patman was also a reliable New Deal supporter. The congressman asked the most fundamental economic question of a democracy: To what extent should the government be responsible for keeping its citizens gainfully employed? Patman's bill would attempt to codify the principles of FDR's Second Bill of Rights speech.

Republicans like the uber conservative Senator Robert Taft of Ohio were skeptical from the beginning, cynically arguing that the legislation creates "a kind of optimism through Government action comparable to the private optimism that led to the 1929 stock market crash." Just as his political descendants would cry inflation to stop any legislation that would raise the standard of living for working families, so too did Taft and his loyal opposition. "If we are going out for the 60,000,000 jobs set forth by Henry Wallace in his recent book of that name," Taft wrote, "I think any effective actions to achieve that objective would bring inflation upon us."[17]

Patman's legislation, whose concept was the brainchild of Wagner and U.S. Senator James Murray,[18] moved through the 79th U.S. Congress and was eventually adopted in a much-watered-down version of the original bill. Notable changes included the removal of specific Keynesian economic goals, the right to have a job and a mandate for the government to maintain consumer purchasing power. Truman signed the bill into law as the Full Employment Act of 1946.[19]

In 1947, as politicians in Washington failed to agree on an economic policy that would benefit working families, the Marshall Plan was kicked into high gear in western Europe. The multinational economic aid program was designed to stop the spread of communism, rebuild political institutions, and promote free-market economies. As U.S. leaders dithered on whether an American had a right to a job, the United States was rebuilding the western European economy by infusing over $150 billion (current dollars) in economic aid between 1948 and 1951.

On the other side of the globe, the former Axis power Japan began to carve its own path to recovery. The United States invested approximately $18 billion (2021 dollars) in the Japanese government and economy.[20] The Japanese flipped the narrative on the United States by using the American-imposed prohibition against rebuilding their military to invest money and resources into more productive, domestic applications like electronics and automotives. In time, they would crush their American competitors.

At about the same time, in 1950, discussions were underway in Paris for the establishment of the European Coal and Steel Community, the precursor to the European Union of 1993. The Treaty of Paris, which established the economic community, would allow for the free flow of steel and coal production among member countries and former World War II belligerents. Its first objective was political: to make another war in Europe not just "unthinkable" but "materially impossible."[21] Its second goal was to convert the region into an economic powerhouse.

Not only was the United States not investing in domestic industry on the same level as the Marshall Plan in the late 1940s but it also lacked a simple industrial strategy to guide tax policy, education and workforce development, trade agreements and investment in basic science research. The United States would navigate the shoals of an emerging global economy rudderless. It would never have a meaningful discussion on this crucial issue, let alone implement an effective, comprehensive policy.

Worse, in 1950, the U.S. government disbanded the U.S. Maritime Commission, which was part of Roosevelt's New Deal legislation. Passed in 1936, this act of Congress subsidized the construction of 5,777 merchant ships over its fourteen-year history.[22] Not only did the U.S. government lack any vision or leadership in building the post-war American economy, but it was also getting rid of the few programs that actually helped industry. (The commission was replaced with the less robust construction differential subsidy program.)

That same year, organized labor committed a self-inflicted wound when Walter Reuther, the progressive United Auto Workers head and co-founder of the left-oriented Congress of Industrial Organization (CIO), signed the so-called Treaty of Detroit with the Big Three automakers (GM, Chrysler, and Ford). In exchange for better wages, pensions, and benefits, organized labor surrendered the right to sit at the table with management to make operational decisions. "It signaled the end of an era in which Reuther could hope for a serious transformation in the governing structures of American industry," according to Lichtenstein. According to *Fortune* magazine, "GM may have paid a billion for peace but it got a bargain. General Motors has regained control over one of the crucial management functions . . . long range scheduling of production, model changes and tool and plant investment."[23]

Corporate America scored an even bigger victory because they eventually got it both ways. Corporations ultimately took complete control of their businesses and the economy as a whole. They poked and prodded politicians into passing pro-corporate/anti-labor laws and policies such as deregulating major industries and implementing supply-side tax cuts. It would be another thirty years before those policies would be enacted, but labor's acquiescence to an anti-labor agenda in 1950 was a crucial turning point. It taught corporate America a lesson on how to accomplish its objective: bribery.

As long as workers had a decent slice of the economic pie, they looked the other way when corporate America signed off on an anti-worker agenda that included free trade deals, regressive tax systems, opposition to a national health insurance plan, and weak workplace safety protections. "Developmental neoliberalism,"[24] as Elizabeth Tandy Shermer named an early variant of the 1970s-style neoliberal economic policy shared by both Democratic and Republican administrations exacerbated income inequality, tempered wage growth, and exploded CEO pay. Today, the average CEO receives 399 times more than the lowest paid worker, and the top 1 percent of income earners hold almost as much wealth as the bottom 90 percent.[25]

By 1952, it was clear that the Marshall Plan was accomplishing its economic goals. Economic output in Europe was well above pre-war levels. All recipient countries saw their output rise by at least 35 percent. Following the plan's expiration in 1951, western Europe continued to experience unprecedented economic growth. Japan was catching up too, achieving economic output parity with pre-war levels in 1956. From 1945 to 1956, the average GDP annual growth in Japan was 7.1 percent.[26]

As Europe and Japan were gaining on the United States, a different foreign competition emerged between the Soviet Union and the United States. The so-called Space Race pitted the two countries' top scientists, some of whom were former Nazi rocket scientists and nuclear physicists divvied up by the superpowers as spoils of war, to see who would enter space first.[27] Although the Russians won round one when Sputnik orbited the earth in 1957, the United States took the main prize when they landed a man on the moon in 1969. Untold technological spin-offs fueled general economic growth in the United States beginning in the 1970s.[28]

Parallel to the space race was an arms race. Shortly after the Korean War (1950–53) both the United States and Soviet Union invested massive amounts of resources into nuclear arsenals. By 1970, the United States and Soviet Union had a combined arsenal of forty thousand nuclear warheads.[29] A larger cache of weapons and missiles may have given one country an advantage militarily over the other, but the nation that could best manage national security and defense expenditures in the overall government budget and economy would ultimately outlast the other.

Union membership and income inequality was stable for the balance of the 1950s in the United States. About 30 percent of national income went to the top 10 percent and union membership sat at 31 percent of the private sector workforce.[30] The initial legacy of the Treaty of Detroit comported with labor's expectations; worker material gains were keeping up with cost of living and benefits remained competitive. Pensions in the private sector were the norm. Or so labor had hoped. Instead of across-the-board gains, islands of "economic security" were carved out through collective bargaining agreements. "Unions, having failed in their ambition to win a national health insurance program, committed to a strategy of negotiating private welfare benefits systems." And only on those private "islands" did workers partake in the fruits of American labor, according to historian Gabriel Winant.[31]

Soon, organized labor would be embedded in the American capitalist structure. The American Federation of Labor (AFL) absorbed the more militant Congress of Industrial Relations (CIO) in 1955 and in place of a thriving labor movement, internationals like the United Auto Workers, United Steelworkers, and International Brotherhood of Teamsters, were content to focus on collective bargaining and financial gains for their members.

In the heart of the industrial rust belt came the country's pioneering right-wing, red baiter senator, Joseph McCarthy of Wisconsin. The Red Scare was not only a political witch hunt of suspected communists in the U.S. government and industry for craven electoral purposes; it had economic objectives too. As Hollywood producers, actors, and writers turned on each other, as McCarthy ended the careers of accomplished public servants in the State Department—and even driving one Senate colleague to suicide when McCarthy threatened to out his son—the American labor movement purged its rank and file of socialists and fellow travelers and moderated union politics and individual union demands.[32] By the 1950s, labor was less radical and increasingly focused on basic material benefits. To sociologist Frances Fox Piven, the chickens of the Wagner Act had finally come home to roost. Just as the National Labor Relations Act brought labor to heel and achieved industrial peace in the throes of the Great Depression, by now the overall movement had cooled to a deep freeze, surrendering any momentum that could trigger anything remotely revolutionary in the labor movement.

According to Piven, the Wagner Act was written to achieve industrial peace amid the labor discord of the 1930s. She contends that it reined in labor activity at a key point when the movement was gaining significant momentum. By codifying key aspects of labor goals, namely union recognition and collective bargaining, the legislation took the air out of the movement's sails and overall activism dried up. The immediate effect was clear. Between 1929 and 1935 the number of striking workers shot up sevenfold from about 200,000 to 1,400,000 on the eve of the enactment of the Wagner Act. Within two years, strikes fell by half, to 800,000. Whether this trend would sustain would be determined later, however.[33] Soon after the passage of the Wagner Act, workers organized the shop floor. The law's framers accomplished their goal: labor peace. By 1940, about one in four private-sector workers belonged to a union.[34]

If there was just one word to describe the 1960s, it was "race." One hundred years after the signing of the Emancipation Proclamation, U.S. race relations continued to divide and incapacitate the country. Lyndon Johnson made civil rights a top legislative priority when he succeeded John F. Kennedy as president in 1963. Johnson, a proud son of the South—a man of great appetites who hungered most for acceptance by the country's political establishment—became, according to contemporary accounts, the first southerner to get to the Oval Office since Andrew Johnson.[35] He would do anything to hold on to power, even if that meant betraying his home region. Johnson fulfilled Kennedy's legacy and broke the heart of Dixie by signing the Civil Rights Act of 1964.

Johnson skillfully capitalized on the nation's mourning period after JFK's assassination and put to use his master-class legislative skill set honed as U.S. Senate majority leader to pass this watershed law. The Civil Rights Act of 1964 is still the cornerstone of civil rights protections for people of color and protected classes

Shortly after passage of civil rights legislation, the president signed the Immigration and Nationality Act of 1965 in New York Harbor under the watchful eye of the

Statue of Liberty. Its intent was to erase laws that unfairly capped immigration from countries outside western Europe. While certain aspects of the law—particularly so-called chain migration and overall lifting of quotas—created downward pressure on wages by flooding the available labor pool, it well complemented Congress's work in eradicating ethnic, racial, and religious bias in American immigration. The 1960s also witnessed the divergence of class- and race-based politics. Since the inception of FDR's New Deal, electoral coalitions on the left had been built around a sense of race-blind class solidarity. Because economic conditions were so stratified and the Democratic Party had strong political leaders like Roosevelt, the class-based coalition hung together until the 1960s. But tears appeared in the fabric of the coalition when racial issues took center stage and—rightfully—politicians started giving more than just lip service to eradicating racial violence and discrimination. It's not surprising that domestic policy shifted from economic issues to racial issues by end of the 1950s with passage of the watershed—though toothless—Civil Rights Act of 1957. Martin Luther King Jr. was passionate about the labor movement and insisted that it was closely intertwined with the civil rights movement. In that respect, it could be argued that the civil rights movement of the late 1950s and 1960s was a natural outgrowth of the economic fights surrounding the New Deal and later anti-union efforts like the Taft-Hartley Act.

"Negroes are almost entirely a working people," King said. "The same politicians who attacked unions were usually the ones who also rejected civil rights."[36] And it was on the occasion of a strike of sanitation workers—many of whom were African American—that King made his fateful trip to Memphis on April 4, 1968.

There was also an electoral incentive. By the time of Minnesota liberal icon Hubert Humphrey's famed address to the Democratic convention of 1948, in which he called on delegates to "get out of the shadow of states' rights and to walk forthrightly into the bright sunshine of human rights,"[37] the Democratic Party had recognized that growing African American populations in cities presented a treasure trove of votes.

Then-presidential candidate John F. Kennedy recognized the same phenomenon in 1960. He skillfully leveraged this emerging Democratic constituency when he made a two-minute phone call to Dr. Martin Luther King Jr.'s wife after the civil rights activist was jailed in Georgia on a trumped-up traffic charge. This gesture did not go unnoticed, nor did his brother and campaign manager Robert Kennedy's call to the sentencing judge to free Dr. King from jail.

According to RFK biographer Larry Tye, it was this act, not any last-minute ballot stuffing in Chicago or the TV debates where Kennedy outshone opponent Richard Nixon, that made the difference in the 1960 contest.[38] In addition to securing the presidency, Kennedy's move marked the end of GOP domination of black politics and the beginning of African Americans becoming a stalwart constituency of the Democratic Party.

This was the culmination of a quarter of a century of dramatic changes in party identification and voter preferences among blacks. Black people had identified as Republicans and largely voted for Republican presidents until the first decades of the twentieth century. African Americans weren't even allowed to attend Democratic national conventions until 1924. In 1936, FDR earned a majority of the black vote (71 percent), and a plurality of African Americans identified as Democrats (45 percent).[39]

Johnson's War on Poverty was the domestic policy equivalent of winning World War II, with the goal of eradicating poverty totally and completely. It was an effort of massive scale.

Johnson was a man of huge appetite, deep insecurity, and high ambition. His sizable electoral mandate in the presidential election of 1964, in which he overwhelmed Arizona Senator Barry Goldwater 61 percent to 39 percent, was also an affirmation of his economic and political philosophies.

Goldwater was the icon of the nascent American conservative movement. He championed an unapologetic right-wing agenda on the campaign trail. Goldwater lambasted Johnson's and the Democrats' progressive economic policies with his call-to-arms line at the 1964 Republican convention, "Extremism in the defense of liberty is no vice." The voters overwhelmingly rejected this platform, and it would be another sixteen years before the country would embrace a Goldwater-like political philosophy and elect a conservative president.

The War on Poverty, like the Vietnam War, is remembered by its ignominious defeat. While its pursuits were noble—employing the unemployable, satiating hunger, nourishing children, protecting elderly and the disadvantaged, educating the masses—it proved too large in scope and, combined with the cost of prosecuting the Vietnam War, way too expensive for the country to afford. Nonetheless, key components of the measures are alive and well today. Most notable among them are Medicaid and Medicare, which form the cornerstone of the nation's health care policy and constitute 23 percent of total federal spending.[40]

For much of the country's history, the environment was a secondary concern. With the notable exception of President Teddy Roosevelt, policymakers were totally focused on economic growth at the expense of the environment. More carbon was emitted into the atmosphere during the industrial period from the eighteenth century to the present than in all of prior human history.[41]

But a boy from Clear Lake, Wisconsin, began to change all that. As Gaylord Nelson (no relation to the author) looked down the main street of his hometown he noticed something missing: trees. Nearby Menominee, Wisconsin, had an archway of tree branches welcoming visitors to their city. Clear Lake didn't. So, the precocious twelve-year-old went to the village board and asked for a similar entrance to the city. They said no. But it wouldn't be the last time they heard from this budding environmentalist.

Eighteen years later and shortly after his return home from his service in World War II, Nelson jumped headfirst into the world of politics. First, he sought the state Assembly seat for Clear Lake as a progressive Republican in 1946 and—after defeat—as a Democrat for a State Senate seat in Madison in 1948.

Nelson would go on to serve a decade in the Wisconsin Senate, two terms as governor (1959–63), and three terms as U.S. senator (1963–1981). By 1970 he had solidified his bona fides as the body's leading authority on environmental issues. Among his accomplishments was founding the Appalachian Trail system, authoring the Wilderness Act, and making Wisconsin's Apostle Islands a national lakeshore. Legislative leaders of both parties sought his advice on conservation topics. Given his stature in this area, it was not surprising he had the audacity to name an entire decade after his pet issue.

The Environmental Decade opened on April 22, 1970, when Nelson established Earth Day, a teach-in across the country to inform people from all walks of life about the importance of conservation efforts and offer a clear understanding of where the environment belonged in the national consciousness: front and center.

Earth Day is celebrated and practiced to this day. The laws that Nelson helped pass in that era include the Clean Air Act of 1970, the Clean Water Act of 1972, the Endangered Species Act of 1973, and the 1970 enabling legislation for the Environmental Protection Agency. This body of law remains the bedrock of U.S. environmental policy today.

These laws were a response to a growing national awareness of industrial pollution and high-profile environmental accidents. Pollution scares were everywhere: oil

(Santa Barbara, California), dioxin (Times Beach, Missouri), toxic chemicals (Love Canal, New York), polychlorinated biphenyls (PCBs; Fox River Valley, Wisconsin), and nuclear radiation (Three Mile Island, Pennsylvania).

As policymakers' attention shifted to environmental matters, a direct result of their failure to properly regulate industry, they lost sight of economic problems. It was not an either-or proposition, with elected officials choosing the environment over jobs and the economy. Gaylord Nelson would have been the first to reject this proposition.

Rather, official Washington overlooked overwhelming evidence of the ills of globalization, including a shrinking manufacturing base, stagnating wages, and downward spiraling union membership.

CHAPTER 3

The Lake They Call Gitche Gumee

U.S. SENATOR GAYLORD NELSON OFTEN REMINDED HIS SENATE COLLEAGUES, "THE economy is a wholly owned subsidiary of the environment."[1] Nelson did so not just on Earth Day but every day because he passionately believed every day ought to be one dedicated to conserving natural resources.

Lake Superior had always been known as Gitche Gumee—meaning "big sea" or "huge water." The continent's Indigenous people, namely the Ojibwe, gave it the moniker simply because no other body measured up to its size and sustenance. Indeed, it is the second largest lake on the planet next to the Caspian Sea; its wild rice beds literally feed a nation; and a fulfilled prophecy that had saved its inhabitants.[2]

Gitche Gumee was the setting for the wreck of the *Edmund Fitzgerald*. On November 10, 1975, a punishing Canadian cold front and an Oklahoma panhandle warm front clashed to produce thrashing waves as high as thirty to thirty-five feet, the so-called three sisters rogue waves that many believe swallowed up the Fitzgerald on its final run.[3] About 550 wrecks lie scattered on the 31,700 square mile lake bed, although only one-half have been discovered.[4] As Gordon Lightfoot's song lyrics tell us, "Superior does not give up her dead." The twenty-nine fallen sailors would learn this admonition the hard way.

Fourteen thousand years ago, glaciers as thick as two miles deep retreated across North America as climates changed and temperatures warmed, scraping deep holes in the earth's crust and forming what would be known as Lake Superior. When these massive sheets of ice melted and water filled cavities in the earth, nutrients accumulated in the way of decayed organisms. Decayed life would add nutrients to the soil from which new life would spring. The cycle persisted and evolved over millennia and continues to this day. But it wasn't just the glaciers that made Superior what it is today. What makes Superior geologically unique to other lakes is its location relative to a failed rift, or fracture, in the earth.

The Keweenaw Rift of Upper Michigan developed over one billion years ago, initially as a hot spot like the present-day Yellowstone in Wyoming. Then, after explosive forces within the earth caused it to "fail," the rift collapsed under its own weight. As it "healed," retreating glaciers carved out the lake basin like an ice cream scoop scraping the bottom of a pail. And just as those last chunks of ice cream were so difficult to get out of the bucket, so too did those massive sheets of ice remain in Superior's pail. Left over in the lake are landmasses like Isle Royale in the very north of Superior's waters or smaller islands near the Michigan side such as Caribou Island, the scene of the crime for the Fitzgerald's sinking.

Lake Superior's uniqueness was not lost on early world travelers. Said one European explorer in 1872: "Those who have never seen Superior get an inadequate, even inaccurate idea, by hearing it spoken of as a 'lake' and to those who have sailed over its vast extent the word sounds ludicrous. Though its waters are fresh and crystal, Superior is a sea. It breeds storms and rains and fogs, like the sea. It is cold in mid-summer as the Atlantic. It is wild, masterful, and dreaded as the Black Sea."[5]

Lake Superior could be considered the *Edmund Fitzgerald* of lakes. Superior is one of the largest lakes in the world. The Fitzgerald was in a class of its own as well. Just as the rest of the Great Lakes relied on Superior to replenish them with new water like one overflowing bathtub spilling into another, America depended on the Fitzgerald to keep the Midwest economy alive and well.[6]

That was all true until the 1970s when the rest of the Great Lakes were overcome by harmful pollutants fed by unkept lakes, particularly Lake Erie, whose Cuyahoga River caught on fire in June of 1969. This is one of several high-profile environmental disasters that built a critical mass of public support for environmental measures that Senator Gaylord Nelson adroitly leveraged to great effect, as described above. It was only a matter of time before Superior also became infected. Said aquatic biologist

Louis G. Williams of the federal National Water Quality Laboratory, a whistle blower who called attention to the declining health of Lake Superior in 1968, "Other writers are now saying that Lake Erie is dead, and that Lake Michigan is sick. Along these lines I would state that Lake Superior is now infected."[7]

Likewise, the Fitzgerald was suffering from chronic maintenance problems by the early 1970s. While some defects were addressed, the structural problems were either patched up or ignored. It was akin to a 1964 Ford Mustang that was not kept up and driven repeatedly over potholes and crumbling roads. The road got worse and eventually the car fell apart.

While boats like the Fitzgerald continued to ply the waters of Gitche Gumee, the community in and around the lake suffered greatly due to severe environmental degradation, particularly the First Peoples who had settled the area fifteen hundred years ago and had stewarded the environment ever since.

The Bad River Tribe of northern Wisconsin, an indigenous tribe of the Anishinaabe or indigenous people of the Great Lakes region, is the area's leading protector of the environment. They have led efforts to halt pipelines, clean up environmental accidents, keep the gray wolf on the protected endangered species list, and ensure the survival of rice, a key staple food for the Anishinaabe and others.

The southern shore of Superior is home to the world's largest coastal wild rice bed. Rice is one of the more water-intensive farming crops.[8] It takes over 800,000 gallons of water to grow one acre of rice.[9] The future of rice as a sustainable food source is dependent on the preservation of its natural habitats, including large spreads like Lake Superior. Wild rice, a "sacred plant" to the Ojibwe, was so important to this Anishinaabe tribe that it was a major factor for their "migration to the Great Lakes from the Eastern Seaboard" according to historian Chantal Norrgard.[10]

Not surprisingly, the Anishinaabe were not appreciated for their contributions toward protecting the environment. Historian Grace Lee Nute, who wrote several books on Great Lakes subjects, including a comprehensive history of Lake Superior itself, went so far to embrace the lake's name and rue the day it would instead honor those who stewarded the waters for centuries. Said Nute, "The English name [Superior] is so apt that it would be a pity for the largest of the Great Lakes to have been named for surrounding Indian tribes."[11]

Nute suggests that if the name were to be changed it should honor the French fur trappers and their transporters (i.e., voyageurs) of the stolen booty who ravaged the countryside from Hudson Bay to Lake Erie. Said Nute, "The Indians of the region call the lake their own, Ojibwa Kitche Gammi (the great lake of the Ojibwa), but it might equally well be termed Voyageur Kitche Gammi."[12]

Notwithstanding Nute's low opinion of the First Peoples—a comment she made in the first edition of her book *Lake Superior* in 1944 and carried over into later editions including the 2000 reprint—to her credit she lets slip the extensive contributions of Native Americans to Lake Superior history: inventing the birch bark canoe that white settlers used as a model for their own rudimentary vessels (though they still made the Indians paddle them around); facilitating and participating in the fur trade; and teaching settlers how to prepare local fare.

Nute also notes the pristine condition Gitche Gumee was in when the white man arrived: "At all events, it is still grand. The great cliffs and rocky headlands are not altered. The gorges and cataracts are as impressive as ever. Fringes of islets in all shapes and forms still add piquancy to the scene. No one has spoiled the crystal purity of Lake Superior's waters, so transparent that small objects may be seen scores of feet below the surface. . . . From Duluth north and east to the Sault [Ste. Marie] the wilderness still prevails; only a few villages and three cities have sprung."[13] That would soon change.

There was a fundamental difference in philosophy about the nature of the land between the Europeans and First Peoples. The white settlers viewed land as capital because the purpose of their journeys was to make money off the land. And those who financed their expeditions expected a nice return on their investment. It was an arrangement that well predated the settlements of the Great Lakes region and could be traced back to the days of Christopher Columbus when the Spanish throne underwrote his voyage to the New World.

Additionally, the Europeans had business partners based in Europe, like the Hudson Bay Company in London. Their loyalty was to those partners, not the natives. Their careless stewardship of the lake and its expansive watershed reflected their true priorities, a far cry from the needs and expectations of the indigenous peoples. No wonder the land would soon be devastated by clear-cutting of forests, mineral extraction, and depletion of fisheries.

The First Peoples had a diametrically opposed land ethic. Most notable was the seventh generation concept developed by the Iroquois, which posited that each generation should treat the environment in such a way that it would be in the same condition seven generations hence.[14] It was no surprise that the continent's resources would be so abundant and land and water so pristine when Europeans arrived in the sixteenth century. The widespread devastation of natural resources and attendant pollution that followed were the work of the unruly interlopers. Evidence of what is called "transgenerational" effects, or exposure to a harmful environmental containment that is inherited by future generations, is well documented. Studies affirm this finding time and again.[15] Seventh generation was not a feel-good philosophy but a crucial policy to stem environmental damage.

America's insatiable appetite for steel, and by extension iron ore, necessary for building up the arsenal of democracy of World War II, depleted iron reserves across the Great Lakes region. As the guns fell silent across Europe and the South Pacific, new national security threats popped up around the globe.

On March 12, 1947, when President Harry Truman asked U.S. Congress for $400 million in aid to Greece and Turkey, both battling a growing communist threat, the Truman Doctrine was born.[16] This new foreign policy pledged sufficient military and economic assistance to countries under threat of authoritarian rule, particularly communism. Thus began the Cold War between the United States and the Soviet Union, which would last for forty-four years, ending on Christmas Day of 1991.

Instantly a massive arms race commenced along with other military and economic competitions, all of which demanded a whole new round of iron ore supplies to build missiles, ships, jets, bombers, rockets, and artillery. Minnesota, Michigan, and Wisconsin together accounted for 85 percent of the country's iron ore supply.[17] If the United States were to win the Cold War or overcome any other future national security threat, it would need the Great Lakes to step up again, and fast. At the center of the drama was Lake Superior.

For the iron ore industry, Lake Superior served as a sort of stock exchange for resource overconsumption and industrial pollution. Companies would draw down mass quantities of water and in return dump tailings—the by-product of taconite ore mining—into the lake. Water in, tailings out.

The leading mining company and consistent offender against state and federal environmental laws, Reserve Mining Company of Silver Bay (the Fitzgerald's upper lake port), drained half a billion gallons of water out of Superior every day. To put this into perspective, all of the communities around the lake who depended on it for water consumed just 5 percent of that total, or twenty-five million gallons.[18]

In return the corporate behemoth dumped 67,000 tons of iron ore tailings *every day* into the lake, or more than twice the cargo capacity of the *Fitzgerald* (26,000 long tons).[19] This was to say nothing of the pollution that would concentrate in these locales and soon spread throughout the massive body of water. And since no one person or entity can own a lake, unlike tracts of land, the water and landfilling were totally free. The scale of this production was breathtaking, yet the amount of iron refined from the facility would meet just *1 percent* of the country's total iron needs.[20]

Such a business model was not sustainable. Environmental and corporate forces clashed in an epic struggle that lasted almost a quarter of century from 1957 to 1980. The otherwise progressive state found ways to stymie federal regulators and make excuses or promote dubious solutions to placate concerned citizens and environmental groups. Not only would the state join forces with corporate polluters to shield them from government regulations but it would also underwrite a good chunk of the company's costs by exempting them from local property taxes. Reserve Mining's tax exemption in turn choked off critical funding to Silver Bay and other municipalities to pay for public safety services like police and fire, and local schools.[21] Such a law was not unique to other states with manufacturing-based economies. Neighboring Wisconsin implemented a tax exemption for all personal property of a manufacturing production facility (i.e., machines and equipment) in the early 1970s. This law came about when the state government was controlled by the Democratic Party, which held the governor's office and majorities in both houses of the legislature.

Policymakers were so transfixed by the economic opportunities of the mining industry that they disregarded its environmental impacts and discounted the economic benefits of a strong tourism industry that contributes $200 million in economic activity each year in the entire state of Minnesota.[22]

It wasn't until 1972 that federal regulators finally took legal action in federal district court. The case of *United States v. Reserve Mining Company* ran a course of eight years from 1972 to 1980. The Environmental Protection Agency, represented by the U.S. Department of Justice, argued that the tailings from taconite production polluted the waters; a substance found in asbestos was discovered in the tailings. By 1974 it was found that tailings had spread over an area of two thousand square miles, a significant chunk of real estate even for the world's largest freshwater lake.[23]

At the conclusion of an expensive and torturous appeals process, Reserve Mining was granted permission to landfill the tailings in an inland pond and not the lake. They continue this practice to this day. It was good news for the lead investor of the mining behemoth who'd had a pretty bad decade: Northwestern Mutual, who fronted 80 percent of the capital construction.[24]

The federal district court case was precedent-setting because it gave the EPA broader oversight authority of corporate polluters, a level of oversight that did not exist previously.[25] But it also pulled back the curtain on the nature of regulation in the United States. Some countries place the burden of proof on the polluter to prove beyond a doubt they do not cause undue harm to the land and human or animal life. In the United States, it is up to the harmed party, the community, to stop the pollution.

It was also an example of how corporations hold jobs "hostage" in communities when called out for polluting the environment. Presiding Federal court judge Miles Lord in the *United States v. Reserve Mining Company* case scolded the defendants for their unbecoming legal and public relations strategy: "In essence, defendants are using the work force as hostages. In order to free the work force at Reserve, the court must permit the continued exposure of [the citizens of Duluth, Minnesota, and other North Shore communities] to known human carcinogens. The court will have no part of this form of economic blackmail."[26]

And neither would the local union who represented the workers have anything to do with it. "The delegates representing all CIO unions in Duluth unanimously went on record opposing the dumping of ore tailings in Lake Superior. We are aware that the dumping of ore tailings into Lake Superior will only establish a precedent for future dumpings [*sic*] of some form of pollution in our beautiful and clear Lake Superior," said a representative of one of the United Steelworkers at a public hearing on the topic.[27]

The union's definition of solidarity went well beyond their own economic interests. This is not the case today, because unions often buy into the either-or proposition: a clean environment or good jobs. But what good is a good job if the worker and his or her family are not healthy? The labor leaders of the greater Duluth and Silver Bay area knew the answer to that question. This is precisely the mindset necessary for present-day labor and industry leaders to rein in the disastrous effects of climate change with a Green New Deal that gears the economy around environmental impacts to keep industry strong, jobs plentiful, and the climate crisis in check.

Compelling examples of devastating fallout from climate change can be found all over the world: wildfires in northern Canada, hurricane after hurricane pounding the southeast United States, hundred-year storms becoming ten-year storms, melting polar ice caps, the inability of species to adapt to new weather patterns, unprecedented flash floods, crop failures, and expanding deserts.

Among the world's lakes, it could be said that Lake Superior has suffered the most.[28] Superior is one of the fastest-warming lakes in the world. Recently surface temperatures have tested fifteen degrees warmer than normal. While the lake measures up to 1,332 feet deep in some parts and thus offers plenty of room for fish and organisms to go deeper as shallow areas warm, fish expose themselves to harm by being closer to predators or farther away from reliable food sources.

Known ice covers have shrunk 79 percent in recent decades. Snow accumulation is down, and precipitation has come more and more in the form of rain. Storms will be expected to pop up with greater frequency and droughts will be more common. Moisture in the soil could decrease, followed by an increase in runoff rates, causing additional soil erosion and further pollution of the lake. This would only compound the problem of the approximately 1.2 trillion gallons of untreated sewage dumped into Lake Superior each year.

Finally, environmental damage due to climate change would have global implications. Boreal forests that cover the northern shore of the lake serve as carbon sinks and capture 30 percent of all land-based carbon worldwide. Rising temperatures will cause carbon to decompose faster in the soil, turning a crucial and otherwise reliable carbon sink into a carbon producer.

Industry, particularly mining, shipping, and steel, may view Superior as simply a working lake, one whose sum value can be quantified in dollars and cents on balance sheets and quarterly profit statements. But that's a shortsighted way to treat an asset and a shameful way to view one of the world's great natural resources.

CHAPTER 4

Rock Bottom

GOVERNMENT CAN GET A BAD RAP. SOMETIMES IT'S WARRANTED. SOMETIMES IT IS not. In the case of the search and rescue attempt by the U.S. Coast Guard to find the *Edmund Fitzgerald*, it was well deserved. The U.S. Coast Guard bungled and botched the effort from start to finish. Woefully unprepared, unequipped, and unqualified, the U.S. Coast Guard—the government entity charged with protecting the country's shores and safeguarding U.S. citizens on inter-national bodies of water—was an unmitigated disaster. An appropriate metaphor would be that it was a wreck.

Everything that could go wrong went wrong. But everything that went wrong was completely avoidable. The Coast Guard was unequipped to respond to the emergency. Their closest cutter, *Woodrush*, was in Duluth, Minnesota. It would have to cover almost the same distance the Fitzgerald traveled over the last twenty-nine hours to get to the scene of the shipwreck. The *Woodrush* would arrive the next evening at 11 p.m., just edging out the Fitzgerald's time, making it to the scene twenty-eight hours after the sinking.[1] The nearest Coast Guard plane was in Traverse City, Michigan, a forty-five-minute flight away. It would take longer to arrive because the crew in Traverse City forgot to load the plane with flares and needed an additional half-hour to load them up. And when they did get to the scene only three flares worked, and they were basically useless. "It cast a yellowish light on the water . . . it was an exercise in futility."[2]

There's more. The equipment that was on hand, the *Naugatuck* in Sault Ste Marie and the *Mackinac* in Cheboygan, Michigan—about halfway between the Soo Locks and Traverse City—were "disabled" and in the shop for repairs. Twenty crew members in Sault Ste Marie would begin fixing the *Naugatuck* post haste. Too little, too late.[3]

Think of it for a moment. The largest lake in the world, renowned for dangerous sailing conditions during this, the worst sailing month of the year, and yet the Coast Guard only had a little plane and a little boat available. And they were nearly six hundred miles apart on opposite ends of the vast body of water. The immediate resources that should have been available were in the shop.

It was the mid-1970s. Budgets were in rough shape, public morale was low, and the U.S. military was struggling to regain its footing in the wake of a brutal, frustrating defeat in a tiny southeast Asian country. As serious as those circumstances may have been, it was little justification for the Coast Guard's pathetic response to the Fitzgerald shipwreck.

"There's a running joke in the merchant marine," Tim Nixon, a long-tenured Great Lakes sailor (1976–1987) and one of Wisconsin's top bankruptcy attorneys, told me. "If you need the Coast Guard, send a post card."[4] This is a cruel but quite fitting descriptor of the government's ineffectual rescue attempt that evening. This laissez-faire approach to the rescue would continue the following day.

On the radio, the Coast Guard response team was not exuding confidence. They sounded like a neighbor leaning on another neighbor to borrow a snow blower to get them out of a fix on a cold, wintry Monday morning.

> **COAST GUARD:** Think there's any possibility that you could turn around do any searching, over?[5]

Arthur M. Anderson skipper Bernie Cooper wasn't champing at the bit to go back out and was grasping for excuses. The hesitancy in Cooper's voice sounded like someone struggling for a good reason not to do something. It reminded me of my late grandfather who would like to lead you to the answer he wanted without asking for it explicitly. "Well, if you really want me to go out and bale the last hay field I will, but church is in two hours, and I promised Rod Peterson I'd help usher." It was a Scandinavian trait that would be handed down to future generations.

> **COOPER:** Oh God, I don't know. That sea out there is tremendously large. If you want me to, I can but I'm not going to be making any time. I'll be lucky to do two or three miles per hour going back out that way, over.

Nevertheless, Cooper's initial response and instinct were understandable. A captain's first responsibility is to his ship and his crew, "and not to a bunch of guys who are dead out in Lake Superior," Nixon stated bluntly. And the storm "was a proven killer," as Nixon put it.[6] As Wisconsin's top bankruptcy lawyer, he wasn't one for mincing words or getting too emotionally attached to his clients.

Of course, Cooper would turn the Anderson around and make the rescue attempt. That was the code of the Great Lakes, the law of the brotherhood. But he did so reluctantly, and he tried having it both ways. He went back, but he slow-rolled the ship. He made effort enough to avoid being called out for neglecting his duty. "So Cooper puts it at 2 miles per hour and heads back out," Nixon explained. The brother of Cooper's first mate Morgan Clark was on the *Philip R. Clarke* that evening and did not approve of this indecision. He felt strongly that the Anderson should have gone back without being asked a second time. For years after, he refused to talk to his brother.[7]

When the Coast Guard at first had no luck persuading Cooper to go back out into the storm, they called up Cooper's boss, Christian Beukema, U.S. Steel's vice president of ore, limestone, and lake shipping. Beukema oversaw U.S. Steel's fleet. He strongly encouraged Cooper to go back. While only the captain can decide, it was, as Nixon termed it, "functionally" an order.

What was truly disconcerting about the conversations between the Anderson and the Fitzgerald and the first piece of evidence that Coast Guard personnel were way behind the eight ball in assessing the situation, became increasing evident as Cooper and the Coast Guard continued talking. It became painfully obvious that the Coast Guard had no idea of the severity of the conditions. Then again, how could they? The weather beacon was down. The Anderson was now their only source of "on-the-ground" weather information. Most other ships were in safe harbor at the direction of the Coast Guard. The radio transcript betrays a clueless U.S. Coast Guard bureaucracy fumbling what would be their highest profile rescue in modern history:

> **COOPER:** Oh, I'll go back and take a look, but God, I'm afraid that I'm gonna take a helluva beating out there on it. If you could send a search ship, if you could get a chopper over there, it'd be blowing over towards Copper Mine point and . . . I

dunno how far in there I can get. *Do you realize what the conditions are out there? You realize what the conditions out there are, don't ya?*

COAST GUARD: This is control. The chopper reports sinking. . . . *We have no other vessels in that vicinity right now*, but it looks like from the information we have it's *fairly certain that the Fitzgerald went down*. And we're talking now of a matter of life or death, looking for survivors of that might be in life rafts, or in the water, and we can only ask the masters to do that without hazarding their vessel. (emphasis added)

That night only the Anderson and Ford answered the call to head back out into the storm to find the Fitzgerald. Incredibly, the William R. Roesch from Oglebay Norton Corporation's fleet, refused entreaties.[8] Others would eventually follow but only after weather conditions improved.

Compared to the *Arthur M. Anderson*, the *William Clay Ford* was a profile in courage. The Ford had taken shelter from the storm and was anchored behind Whitefish Point. The Coast Guard had strongly recommended ships take safe harbor. The Ford, while similar in size to the Fitzgerald, was riding light because it was upbound with no cargo—just ballast—and did not have adequate buoyance for the rough winter storm waters.

After the Coast Guard asked Don "Windy" Erickson to go back out into the storm, Erickson thought briefly about it, then picked up anchor and headed out in search of the Fitzgerald. Erickson and the crew scoured the area for six hours and only left when they were relieved. "We tried. We gave it our best," Erickson said, which was more than anyone else on the waters that night could say.[9]

Some members of the maritime community thought it was foolish and irresponsible to go out in that weather. But Erickson had his reasons, and they were straightforward. And honorable. "That could have been my son, my brother, and I would want somebody to go look," Erickson said, according to Nixon.[10]

The night search was an exercise in futility. Neither the Anderson nor the Ford nor aircraft spotted any remnants of the ship. Not until morning did the Anderson find any evidence of the wreckage. Lifeboat number one was discovered about nine miles from where the wreckage would ultimately be found. It looked like an angry seamstress ripped the boat apart along the bow of the keel.[11]

Later that day, the Coast Guard spotted "a spreading oil slick bubbling up from the frigid depths of the storm swept lake."[12] It was the first tangible piece of evidence that directed rescuers to the approximate location of where the ship sat underwater.

Two days later, on Thursday, November 13, a P3 Navy plane deployed in the area identified a large structure on the lake bottom about 1.8 miles from the Fitzgerald's last sighting (via radar), though officials were not entirely positive it was the Fitzgerald.[13] On Sunday, November 16, 1975, the Coast Guard claimed they "detected the wreckage under 535 feet of water" and were "'pretty certain' it was from the Fitzgerald."[14] But it wasn't until November 27, nearly two and a half weeks after the ship went down, that the wreckage in question was positively identified as the Fitzgerald.[15]

While they may not have risen to the occasion in the search and rescue effort, the Coast Guard at least understood the fierce urgency of the moment. The families demanded and deserved answers as to how and why the ship sank and whether they'd ever be able to retrieve the bodies of their loved ones. "The most important task facing the Coast Guard is trying to locate the wreckage," Lieutenant William Holt explained. "Then we could determine what happened. If it was something that could be corrected, then it's important that we see that this is done by other vessels."[16]

For the families of the Fitzgerald crew, it would be another six months before a dive could be made and begin offering clues to why and how the Fitz sank. Citing "weather conditions" and "because the wreckage was lying on the bottom of Lake Superior in 530 feet of water," the government would not undertake a "comprehensive examination" of the wreckage until May of 1976.[17] Leading Fitzgerald historian Fred Stonehouse never accepted the Coast Guard's argument and offered a scathing critique of this decision, as well as the Coast Guard's initial search and rescue effort, in his compendium *The Wreck of the Edmund Fitzgerald*: "The Coast Guard waited until spring of 1976, before undertaking photographic surveys of the Fitzgerald wreckage. This seems an inordinately long period. There was apparently no greater sense of urgency in determining the possible cause of her loss than there had been for accurately assessing the situation and taking immediate, appropriate action on the night she was lost."[18]

To conduct the "examination," the examiners would borrow the U.S. Navy's CURV III System, an unmanned submersible vehicle that goes into deep waters, often under imperfect conditions, to capture video and pictures of wreckage. It also tracks nuclear weapons

On January 17, 1966, the U.S. Air Force lost a thermonuclear weapon off the Mediterranean coast. An armed B-52 collided with a refuel tanker in midair and lost its entire payload. The U.S. military quickly assembled an all-star search and rescue team and got their hands on plenty of state-of-the-art equipment, including the CURV I, a predecessor of CURV III, to locate and retrieve the errant bomb. The search commenced immediately. The bomb was discovered four months later, 2,900 feet under water.[19] The Fitzgerald families were not as fortunate as the U.S. Air Force.

The Coast Guard knew where the ship was, it was in much shallower depths than the H-bomb (one-sixth the depth), it had newer equipment available, and the wreckage was safer to dive than an area surrounding an undetonated warhead. But despite these advantages, they did not make an effort to conduct the search for six months. Companies like Oglebay Norton Corporation and U.S. Steel run their ships until the middle of December. The Coast Guard was unable to drop an unmanned submarine at 530 feet to take some pictures. Would the government have acted more quickly if the *Edmund Fitzgerald* was sitting on an undetonated warhead?

In May of 1976, the Coast Guard ramped up their effort to overkill, conducting a nine-day marathon dive with twelve trips, yielding forty-five thousand feet of videotape and nearly one thousand photographs. The twelfth dive must have felt like the ninth trip to the moon to collect even more moon rocks.

Meanwhile, the Fitzgerald families had suffered through six months of agonizing waiting. Widows were trying to figure out how to put food on the table and a roof over their families' heads. Young children like Michael Armagost's Christopher and Michele were without dads. Insurance companies were giving families the runaround. Oglebay Norton Corporation and Northwestern Mutual were threatening witnesses at the Coast Guard investigation committee. And families were no closer to learning how and why their loved ones perished.

Hours after the ship went down, two theories emerged as to why the ship sank: (1) the hatches on the deck failed and broke open, taking on too much later and sinking the ship; and (2) the 729-foot freighter was snapped in two by massive, successive waves. The latter would evolve into the Three Sisters theory.

The Coast Guard put forward a variation of the second theory, contending the Fitzgerald "probably snapped in two" because it was "heavily loaded."[20] The government entity that would be conducting the official investigation into the sinking should not have proffered opinions on the matter beforehand.

There was more. The Coast Guard proceeded to stick its nose into Captain Cooper's business by not so subtly calling him out for making the wise initial decision not to speak to media or comment publicly on the wreck. "I don't know why the captain of the Anderson hasn't made a statement," said Chief Jere Bennett of the Coast Guard.

Instead, Cooper spent most of November 11 huddled with executives from U.S. Steel and Oglebay Norton Corporation in Sault Ste Marie. If you were the witness to a crime that took 29 lives, you'd want to talk to a lawyer first.[21] It is, however, noteworthy that one of the first people Cooper met with was Renold Thompson, the Oglebay Norton Corporation executive who saw with his own eyes the Fitz's damaged hull (see chapter 6). One can only imagine how that conversation began and how thick the nondisclosure agreement contract was.

Then the Coast Guard one-upped themselves by offering legal commentary on Cooper's actions. "The company may be reluctant to have him talk because of litigation that might come of this through next of kin of missing crewmen," Captain Charles Millradt of the U.S. Coast Guard Soo Locks station.[22]

The Coast Guard wasn't any easier on the families. Bennett publicly kited the possibility that crew members were "trapped," evoking an unthinkable fate that would only serve to further torture the devastated wives, sons, and daughters. "The temperature of the water has a lot to do with it as well as whether or not they've been trapped. Generally, the warmer the water, the quicker they come up."[23] If it were a possibility that crew members were trapped, the Coast Guard could have handled it with better tact, or pull a Bernie Cooper and say, "no comment."[24]

On Friday, November 14, Admiral Adam Owen W. Siler, the commandant of the U.S. Coast Guard in Washington, DC, ordered an inquiry into the sinking of the Fitzgerald. A five-member board was established that would be headed by Rear Admiral Winford W. Barrow, a well-respected official. Board members were also named, and the first meeting was scheduled for Tuesday, November 18, eight days after the Fitz's sinking.

The inquiry's purpose, according to the commandant, would be "to attempt to determine cause of the sinking and whether any new regulations might be needed to prevent similar tragedies in the future."[25] No Coast Guard official walked back any comments made the previous week that would color the investigation. It was one thing to speak out of school in the heat of the moment, another to not admit one's mistakes and carry forth as if nothing happened. Then again, the past week had been a blur. The shipping industry's equivalent of the sinking of the *Titanic* had occurred. One of the largest ships in the Great Lakes fleet had gone down, one that many thought was unsinkable. As the Coast Guard continued their missteps, it was clear matters were not going to improve. The Guard was not up for the task, and families and the public were slowly losing confidence. They needed a champion.

Attorney Vic Hanson, general counsel for the Seafarers International Union, quickly seized the moment and spoke up for the families.[26] Hanson and his co-counsel Nel L. Mann represented Karren Pratt, widow of Second Mate James Pratt, and Mary Poviach, widow of Wheelsman John Poviach. On Monday November 17, Mann filed a lawsuit in U.S. District Court on behalf of Pratt and Poviach and against Oglebay Norton Corporation for sailing an "unseaworthy" vessel.[27]

Hanson was no stranger to the limelight, particularly in the wake of high-profile ship sinkings like that of the Fitzgerald. Hanson had carved out a niche for himself as a plaintiff's attorney in recent ship sinkings, including the *Carl D. Bradley* (1958), *Cedarville* (1965), and *Daniel J. Morrell* (1966). In his first case, the Bradley, he won a record judgement for the families against U.S. Steel to the tune of $1.25 million ($13 million in 2023).[28] In the shipping community, Hanson's reputation preceded him. So, when Oglebay Norton Corporation heard Hanson was on the case and beginning to sign up clients, they had cause for concern. The sinking of the Fitzgerald would not be written off as a freak accident. Someone was going to pay.

Hanson pushed hard on the idea that the Coast Guard board should be disqualified from conducting the inquiry because of a conflict of interest. Hanson said, "It took too long for the Coast Guard to respond to the disaster and the board of inquiry should not sit in judgement of its own conduct in the Fitzgerald disaster." Hanson "planned to challenge the board of inquiry on possible negligence in the Coast Guard's handling of rescue efforts."[29]

On November 18, attorneys for the United Steelworkers of America (USW) Local 5000, which represented twenty crew members and the Marine Engineers Beneficial Association (MEBA), which represented eight officers, petitioned the Coast Guard to intervene as "parties at interest," which would allow them to take on a more active and participatory role in the investigation. The Coast Guard demurred because they did not face liability. Rather, Oglebay Norton Corporation and the late Captain McSorley were parties of interest because they were the subjects of the investigation since one owned the ship and the other steered it.[30] The USW and MEBA would only be allowed to submit questions to the board.

Tellingly, Northwestern Mutual was not named a subject of the investigation even though they owned the vessel and had commissioned it, casting further doubt on the Coast Guard's competence. Any matter of unseaworthiness could be traced back to its original design and construction. And this was done by Northwestern Mutual—not Oglebay Norton Corporation. Then again, the insurance giant had signed a triple net lease with Oglebay Norton Corporation absolving them of any liability. They more than any other company would know how to structure a future insurance claim to save its hide.

In retrospect, it was a setup from the get-go. The purpose of the inquiry was to protect governmental and private institutions. The Associated Press correctly identified the bureaucratic breakneck speed of the Coast Guard as a peremptory move to protect industry in general and Oglebay Norton Corporation in particular from legal responsibility toward the crew—as well as the good name of the U.S. Coast Guard itself. "While the board is not empowered to fix criminal or civil liability, both attorney Nel Mann [co-counsel for two families] and Coast Guard spokesmen agreed its findings may have a bearing on an anticipated rash of suits filed by survivors of the 29 crewmen lost on the Fitzgerald."[31]

Assuming the news account accurately reported the sentiments of the Coast Guard and the attorney for the two families, it was insulting beyond the pale to refer to justice for working families whose loved ones may have died needlessly as the "displaying or proceeding from a lack of careful consideration of the possible consequences of an action."

Because the Coast Guard investigation would not be objective and liability would not be assigned, the families would have to press forward with weak cases and likely move to settlement talks with the company and its insurers. Some families would not even seek legal assistance. Sometimes their calls to the company were not even returned. One month after the sinking, on December 3, 1975, Oglebay Norton Corporation president John J. Dwyer told the United Press International, "We all mourn the shocking loss of the Fitzgerald and her crew. Oglebay Norton Corporation is in close touch with every family involved to make sure that none lacks the immediate financial support it needs."[32] If they were in fact in touch, it was to settle, sign nondisclosure agreements, and get rid of them.

In early December, the Coast Guard inquiry was moving along and making progress. They were on track to finish up testimony later that month. The board members anticipated spending the rest of the winter studying and reviewing the transcripts and collected material.

Nevertheless, they did hit a few bumps in the road. While the Guard was ascertaining the location of the wreckage, the board was trying to track down Cedric Woodward, pilot for the Swedish-flagged saltwater ship the *Avafors*, whom they considered “a key witness.” “He was probably as close to the scene as any other witness, someone who perhaps could have seen the lights of the Fitzgerald,” said board chairman Barrow. Woodward, who refused to go back out into the storm to find the Fitzgerald on November 10, could not be tracked down. Reports had him hunting somewhere in the Upper Peninsula.[33] It seemed incredible that a major government investigation stalled because a key witness could not be found. Then again, it was hunting season in Michigan and a time well before mobile phones. Hunters made sure that when they went on the annual hunt, they could not be bothered or found. By anyone. Including the U.S. government.

When Woodward did surface, board members yawned. His testimony emphasized the Six Fathom Shoal. Woodward was certain that the ship had bottomed out and had been collecting water for three and a half hours. Combined with high seas, she ultimately succumbed to mother nature and sank. The Coast Guard referenced Woodward’s testimony, but ultimately discounted his Six Fathom Shoal theory in its final report.

Before the board made much headway, the United Steelworkers of America made their opinions on the sinking known. Appalled at the industry’s utter lack of care and concern for the wellbeing of workers, one of the country’s largest industrial unions repeated its call to mandate waterproof bulkheads in the cargo hold. Making bulkheads impermeable would prevent cargo breaking through walls in the hold and sloshing together and offsetting the ship’s center of gravity, causing structural damage or sinkings.

The USW called attention to an important point that others seemed to overlook. Why did McSorley tell the Anderson he was holding his own when he knew that the boat had been taking on water and had a list for three and a half hours?[34] Was he lying? Or did this forty-four-year veteran of the lakes not put two and two together and realize that taking on water plus listing equaled disaster and could sink the ship?

And if McSorley knew his ship was in trouble, why didn’t he seek safe harbor? The USW wondered aloud in its December 1975 monthly newsletter *Steel News*. USW suggested McSorley could have dropped anchor near “Isle Royale, the leaf of Michipicoten or Michipicoten Bay.”[35] While that may be true, the ship had passed those locations before McSorley was aware he was taking on a list and water. True, he had already been notified of severe weather storms, but if ore traffic travel would stop after every serious storm warning issued on the lakes, the United States steel industry would grind to a halt.

At the end of the day, it was all for naught. All the witnesses, all the testimony, all the data, all the time, all the attention, all the money ($1 million) added up to nothing. The National Transportation Safety Board, the government agency charged with overseeing transport-related accidents and making recommendations for safety improvements, rejected the Coast Guard report out of hand. It did so unanimously and within two minutes of evaluating the report.[36] The government was eating its own.

The National Transportation Safety Board did issue its own set of recommendations. Many of them were implemented, including the mandating of exposure suits to prevent hypothermia in cold weather conditions, establishing emergency beacons, and implementing a one-compartment stability standard that would limit flooding in the cargo hold.[37] Nevertheless, the wheels of government moved slowly. The Coast Guard wasn't the only federal agency to bungle things. Improvements were not made until well after the sinking, and it took eleven shipping seasons after the Fitzgerald went down for all three key safety measures to be put into effect.

The only other inquiry on the sinking was conducted by a special interest group. The Lake Carriers Association report was intended as a fig leaf to provide cover for Oglebay Norton Corporation and the shipping industry. Its conclusions were typical lobbyist fare—"when changes running into millions of dollars are recommended based on a possible cause of an accident, industry most vigorously objects"[38]—and was not taken seriously.

It seemed the only thing that these diverging interests could agree on was that the Coast Guard report was bunk. No one had anything good to say about it. This was ironic, because the final report was a political document through and through and was meant to make as many people happy as possible. But just as Aesop warns in his fable "The Miller, His Son and the Ass," when you try to be all things to everyone, you can end up pleasing no one. Or worse, end up being laughed at by everyone.

Perhaps everyone would be better off had the Coast Guard report never been written in the first place. The final document had the effect of not just confusing readers in general but sowing confusion as to who was actually responsible for the sinking and, most important, whether the ship was seaworthy. The outcomes of lawsuits and insurance settlements rested on the answer to that last question.

The report assigned blame to faulty hatches. But once again, the report was opaque. It wasn't clear if the hatches failed because they were manufactured incorrectly, maintained poorly, or fastened incorrectly. When the report stated, "There is no evidence of actionable misconduct, inattention to duty, negligence or willful violation of law," was it referring to the captain, the crew, the owner, or all three?[39] To state the tragically obvious, the board called for "an increased awareness of the hazards that exist [on the Great Lakes]."[40] You can't make this up.

At the end of the day, the message the U.S. Coast Guard sent to the shipping industry and the American public had less bite than a schoolteacher admonishing horseplay on a children's playground. For now, the families would have to take Oglebay Norton Corporation's word and hope they'd get a phone call, a lifeline, or a little help. That's all they had. For now.

CHAPTER 5

Just Call Toby

THE FAMILIES OF THE SAILORS OF THE *EDMUND FITZGERALD* DO NOT LIKE TO TALK about their insurance settlements with Oglebay Norton Corporation. "It's a sore subject," Deborah Champeau, daughter of Third Assistant Engineer Oliver Champeau told me.[1]

And for good reason. Despite their fathers, husbands, and sons being in the same boat—literally and figuratively—the settlement amounts were as wide ranging as the random, roaming waves of Lake Superior itself. Of the family members and attorneys I spoke with, settlements ranged from $25,000 ($122,000 in 2023) to almost half a million dollars ($2.4 million in 2023). While inherent factors such as age and earning potential figured high in calculations for settlements, the range of payments were well beyond any disparities due to these considerations.[2]

Payments were settled between a few months after the ship's sinking and as late as 1982. By the latter date the country was facing its worst levels of inflation in thirty years. In the six-year period between 1976 and 1982, inflation shot up a staggering 76 percent. Payouts in 1982 would thus have had to be nearly twice as much to match the settlement amounts made at the beginning of the ordeal.[3] Earlier in the decade as momentum was gaining for laws to protect natural resources and keep communities healthy and workplace protections through the Occupational

Safety and Health Act (OSHA), the movement for tort law reform with this express intention of discouraging lawsuits against negligent corporations—perhaps a Great Lakes freight shipping company—was building steam. The modern-day tort reform movement began in the 1970s, a campaign that was largely underwritten by corporations, particularly insurance companies. Insurance companies were not interested in torpedoing claims one by one—whole floors of claims agents took care of those nuisances—but by overhauling "public perceptions" and pushing hard for "legislation limiting personal injury lawsuits."[4] Consumer advocate Ralph Nader—who earned national acclaim for his book *Unsafe at Any Speed*, which exposed the auto industry's shoddy safety record—was in full gear pushing back on tort reform movement, earning him the moniker from opponents as a "one-man cheering squad for tort law" as if he and he alone would benefit from consumer protections.[5] His legacy would continue well into the next century.

Families could be sorted into two groups: those with legal representation and those without. Given the modest backgrounds of the families who had little to no access to high-powered legal expertise, most struck out on their own. Oglebay Norton Corporation exploited this reality to maximum effect. The disparate payouts proved it. Those without representation settled quickly and signed nondisclosure agreements. They received considerably less than widows and other surviving family members who had representation.

As soon as word got out that some families received more than others, those without attorneys sought out legal representation. But it was too late. These family members were turned away from law firms because they had signed agreements with Oglebay Norton Corporation and/or Northwestern Mutual that precluded them from recovering any more damages.[6] This was also the case for families that worked with attorneys with limited maritime law experience, who signed off on relatively small settlements.

Maritime law is a two-hundred-year-old body of law that is weighted against workers and consumers. In the nineteenth and even eighteenth centuries, the fledgling U.S. government, seeking to encourage investment in trade, imposed legal limits on carrier liability. And while laws that govern other industries like rail have evolved as the inherent dangers of the work have been recognized, maritime law has not changed one iota.[7]

Unlike in the rail, steel, or paper industries, when workers are hurt—or in the case of the sinking of the *Fitzgerald*, perish—victims and their families must meet a high burden of proof before companies must take responsibility and pay up for damages.

The fundamental question of responsibility is whether a sinking was due to negligence on the part of the captain or crew or whether it was due to unseaworthiness of the ship. If the vessel sinks because of negligence, damages are limited to the value of the ship and cargo, which are at the bottom of a lake or ocean. British common law, the basis of American maritime law and all U.S. federal law, pegged the value of a sunken ship at "two pence," the equivalent of two American cents. For the purposes of this case, however, Oglebay Norton Corporation and Northwestern Mutual were a bit more generous and pegged its value at $817,920.[8] However, if the ship sinks because it was not seaworthy, and should never have sailed in the first place, the damages are unlimited.[9]

It was this question that drove early inquiries surrounding the nature of the Fitz's demise. It remains a compelling question among history buffs, divers, maritime enthusiasts, and family members who are consumed by the tragedy. It is a big reason for the enduring legacy and mystery surrounding the wreck. But among lawyers involved in the cases it quickly became a moot point, for the simple reason that neither side wanted to take the case to court.

For Oglebay Norton Corporation, theirs would be a tough case to defend in court, and they knew it. The facts spoke for themselves. A ship sank and twenty-nine sailors died. In the legal world it was considered *res ipsa loquitur* (the thing speaks for itself) or, in lay terms, it was a slam dunk.

Northwestern Mutual wanted their name out of the newspapers. (It didn't help that the boat was named after a company CEO.) They had customers to keep happy. It wasn't good for business if an insurance company was part of a high-profile story in which policy holders on deceased victims were being denied or cheated on legitimate claims. At first, Haskell seemed to recall, Northwestern Mutual would not even acknowledge to the families that their ship had sunk. "Everybody was saying, 'Okay, where is it then?' If it isn't their ship that went down, where is their ship?" James Haskell, little brother of Second Assistant Engineer Russell Haskell, told me. "And three or four days later they said, yep it was the Fitzgerald that went down."[10]

Oglebay Norton Corporation was also slow to inform the families of the tragedy. At the same time, they were getting families to accept insultingly low payouts and sign gag orders. The company was desperate to save face and money, even though it had posted record profits for 1975.

The families wanted the whole matter put to rest so they could *try* to get on with their lives. Plus, they could use the money. For most of these families, their loved one who died was the household's only breadwinner. "Breadwinner liberalism," as Robert O. Self named it, developed in the nineteenth century. This way of life segmented out male and female household roles. Men's work was "remunerative, family sustaining," while the female was the "caregiver" typically without remunerative work.[11] In other words, the best a sailor's family could hope for in the 1970s was a well-funded nineteenth-century family model. Families immediately found themselves without an income stream, and good paychecks in the Northwoods were hard to come by.

Healthy paychecks were so rare in Superior and a large swath of northern Wisconsin that when sailors married young, jealousy would abound among their former dependents, such as siblings or parents. Fights between wives and in-laws over money would ensue. An already harried family life where fathers, husbands, and sons were away from home eight months out of the year, missing birthdays, graduations, anniversaries, and holidays, would destabilize home life even more. The families needed a champion.

If you could distill the American dream into one person, it would be Toby Marcovich. In a sea of Scandinavians and Lutherans, this Jew of Russian and Lithuanian extraction stood out in Superior. "Well maybe not stood, I'm only five feet seven," he told me.[12] Toby was not immune to his community's dry, self-deprecating Norwegian humor. Said Steve Eckman, an attorney of the DeParque Law of Minneapolis, Minnesota, a firm that represented two Fitzgerald families, "He was flamboyant, he was unorthodox. He was a bit like the show on TV, *Better Call Saul.* He was like [lead character] Jimmy McGill. That's Toby."[13]

The ninety-two-year-old attorney sat across from me in his law office, behind a full-size executive desk cluttered with legal memos, knickknacks, and coffee mugs. His permanent and sincere smile made me feel like a close friend even though we had just met in person. No wonder he was such a successful attorney. In addition to black-and-white pictures marking career and legal milestones placed on the wood-paneled walls of his office, a sundrenched photo of Marcovich and his wife atop a Bradley tank with the Israeli army hung next to the door. It was the first and last thing visitors, including opposing counsel, would notice when entering and leaving his office.

More important, Toby stood out because of his legal acumen in the courtroom. In the early 1970s Toby was winning cases across the legal spectrum: settling personal injury cases for record amounts, getting small-time crooks out of murder convictions, and representing families of the fallen sailors of the *Edmund Fitzgerald.*

His connection to the families of the Fitzgerald was natural, and not just because he was the top plaintiff's attorney in the Northwoods. From the docks of Superior and Duluth, Minnesota, to the imposing shipyards of Ohio, Toby called the USS *Philip Minch* home for the summers of 1947, 1948, and part of 1949. He was one of them.

Toby was a local boy made good. He earned his JD from the University of Wisconsin–Madison in 1954. He turned down a professorship at Stanford University to return home, where he was elected Superior city attorney, later served as an assistant district attorney, and finally established a private practice in 1968.

You could say Toby was a renaissance man of sorts. An amateur pilot, Toby would log 5,000 air hours over his life. In his single-engine Cessna airplane, which translated into over 700,000 miles or twenty-eight laps around the planet.[14] A good chunk of those miles was logged between the city airport in Superior and Chicago's O'Hare International Airport when he was settling cases for two of his three Fitzgerald clients, Janice Armagost and Doreen Cundy. The third client was the family of Porter Freddie Beecher. Beecher was single with no dependents. Maritime law considers a life worth less if the soul had an uncle and not a spouse or children, as was the case for Mr. Beecher. And even that didn't mean much to Oglebay Norton Corporation. "I was able to settle it for a small amount just so they'd get rid of me," Toby told me.

Toby also had close connections to the families. Beecher's uncle was a cop in Superior and Toby worked closely with him when he was a prosecuting attorney for the county. Toby had a couple of connections to Ransom Cundy, Beecher's best friend. Cundy's wife was a waitress at a nearby diner that Toby frequented. Cundy was also a good customer at the corner bar, and Toby knew the owner. Toby wrote a will and last testament for Michael Armagost. He would run into the Armagosts for Friday night fish fries at a supper club in Iron River, thirty-eight miles east of Superior, near where Toby had a cabin.

Toby found Mrs. Armagost in his office waiting room the morning after the ship's sinking. She knew exactly what to do when word of the ship's demise reached her. Before setting off to the Superior docks just two days earlier, Michael had told his wife, "If anything happens to me, just call Toby."

It was in a Chicago boardroom where Toby went head-to-head with a phalanx of lawyers defending the good name and reputation of their client, Oglebay Norton Corporation. But Toby would not be intimidated by the big city attorneys. From an early age, he learned to fight back against schoolyard bullies who tried to take advantage of his size and toss around cruel jokes. He endured those attacks and insults for some time. "That was, until I learned to fight back," Toby told me. Now, attorney Toby would need that kind of fortitude and endurance for the Fitzgerald cases. His first trip to the Windy City on his Cessna 337 would not be his last.

Overall, the attorneys for the families of the Fitz didn't have much to go on. Their leading piece of evidence was at the bottom of a lake, covered with and surrounded by 416 million taconite pellets. Plus, no one knew what they would find down there—or when. Photographic and film technology was relatively rudimentary at the time, and no one was confident in its quality. There could be a smoking gun, or there could be nothing. No one wanted to wait around any longer. Justice moved slowly enough.

Coast Guard investigators put eyeballs on the wreckage in May of 1976.[15] Not surprisingly, the pictures showed a ship battered and bruised and snapped in half with twenty-six thousand long tons of taconite pellets scattered on and around the wreckage. There were no bodies nor any compelling pieces of evidence. The families and loved ones of the fallen sailors weren't any closer to the answers to the important questions: why it sank, what caused it, and who was responsible. Families and their attorneys would have to look elsewhere or hope for a bit of luck to come their way.

For Toby, luck came early. Incredibly, Ransom Cundy had taken a picture of his wife on top of the ballast tanks in the hull of the Fitzgerald a few years earlier. The ballast tank is at the bottom of the inside of a ship's hull. It is about six feet deep. The area is filled with water when ships make their return trip to port without cargo. Its purpose is to give the ship stability as she sails. At first blush, it seemed completely ridiculous. "Why the hell anybody would go down there and take a picture is beyond me," Toby recalled. "She's standing on top of the ballast tank in front of a rib. It's just nuts. Getting down there is a project in itself."

Crawling down into a ballast tank is not unlike slipping down a manhole in the middle of a street. There is a lot of activity upstairs, but under the street—behind the scenes—hides the true anatomy that makes the road work. Downstairs it's ugly and raw: unkept and rusted stairs, pitch-black surroundings, and the constant ringing of noises above and below the hold. If a sailor is inside the ballast area for too long, he could fall victim to dead air and pass out. Work in the ballast was considered the least desirable of chores on a ship.[16]

The Fitzgerald was known by reputation as a "wet ship"[17] because the ballast never completely emptied from the tanks. It would collect lake-bottom dirt and sludge, which would accumulate and measure up to three feet high at times. Worse, the crew would only clean it out when ordered by the Coast Guard, and even then it was done half-way. Over time, collected water and dirt in the hull could rust the ship's bottom and weaken its structure.

What was truly incredible about Cundy's picture is that it showed a hairline crack in a rib on the hull. There she was in all her glory—the smoking gun that would close the case. While slightly faded, the crack ever so carefully crawled down the side behind Mrs. Cundy's right shoulder. It was real. Toby's legal career reflected a charmed life, and this encounter was no exception. He sent the picture off to a lab to be analyzed.

One report was inconclusive, but the other one affirmed his initial suspicions. He sent the latter ("Of course, I only sent the one that helped my case.") to the plaintiffs' attorneys committee, a loose consortium of lawyers representing various families. Despite the expert opinion, Toby's colleagues seemed unimpressed. But for Toby it did come in handy. He held onto the photo until the very end of his negotiations. When he disclosed the evidence, the attorney settled soon after. Following the conclusion of the negotiations, he filed it away in a banker's box in his law office basement where it would then be lost to an accidental flooding several years later.

The unresponsiveness of the committee was likely due to the timeline of the case. The committee was in the process of disbanding around late 1979 and early 1980, according to Toby, with each attorney negotiating a separate settlement for their clients. Some attorneys and families had already settled. Each attorney would receive a healthy contingency fee that they wouldn't have to share with anyone else. Debates about whether the sinking was due to negligence or unseaworthiness were long gone. Now, it came down to legal—and political—acumen. Toby checked both boxes.

During the course of negotiations on the Cundy and Armagost cases, the lead attorney for Oglebay Norton Corporation was diagnosed with malignant cancer. Week after week, trip after trip, the defense attorney appeared weaker and weaker to Toby. It was an opportune time for Toby to take full advantage of the attorney's declining health, to wait it out as long as possible. But Toby didn't. He didn't have to.

Evidently, settling insurance cases was high on the defense attorney's bucket list. In the middle of 1981, Oglebay Norton Corporation's lawyer confided in Toby that settlement was a priority for him personally and not just his client. "I think the Fitz was a different kind of case," Toby told me. "We've had a lot of ships sink. It just didn't create the kind of aura like the Fitz did. The Fitz was just different."

Both Toby and the attorney representing Oglebay Norton Corporation redoubled their efforts to resolve the stalemate. This working-class hero—this man of humble origins who never forgot where he came from—did right by his clients. In early 1982, over six years after the *Edmund Fitzgerald* sank, he got "82 percent of the maximum" ($250,000) for Janice Armagost and slightly less for Mrs. Cundy. The formula for damages that was used, according to Toby, was based on precedent set by the case *Cederberg v. Robison* from 1893, decided by the California Supreme Court.[18]

But that wasn't the end. His luck had not yet run its course, and he was on a roll. He was able to take advantage of the devastating economic times, as double-digit inflation rates were strangling the economy in 1982. "At the very end, when they finally agreed on the settlement. I got a call from one of the Lloyds [of London] guys, one of the managers," Toby told me, referring to Oglebay Norton Corporation's overage insurer—a consortium of insurance companies and investors put together by Lloyds of London—who paid all claims beyond Oglebay's approximately $1 million deductible. "He said, 'Would you be interested in deferring your payments for six months, if we pay you 20 percent interest?' And I said, 'Well, I'd recommend it to my clients.'" Both families quickly agreed. Lloyds of London saved their investors 24 percent on interest and Toby's clients received thousands more off the principal, more than some of the families received from insurance companies individually.

While the Cundys and Armagosts received some justice with their settlements, the matter of what really happened, why it happened, and where responsibility lay continued to elude the families. They demanded answers and they deserved the truth. And in time it would come, thanks to another proud son of Wisconsin's Northwoods.

"Golf is my vocation," Dick Theno told me across the table at a Perkins restaurant in suburban Saint Paul.[19] I strained to make out the words; I could barely hear him over the din of pots and pans clanging in the kitchen, which were competing with the voices of hungry kids begging their parents for Mickey Mouse pancakes. But I had waited a long time to meet him, and I wanted to hear every word.

Dick Theno was the retired investigator for Dick Hunegs's law firm, which represented the families of Allen Kalmon and Blaine Wilhelm. There are but a handful of people left connected to the *Edmund Fitzgerald* who hold the keys that can unlock the mysteries behind its demise and sinking. Dick Theno is one of them.

The Ashland, Wisconsin, native is careful to use the term "demise" and not simply "sinking" when he speaks about the Fitz. He knows her fate was sealed many years before November 10, 1975, and he can prove it. He found himself helping two of the families of the fallen sailors of the Fitzgerald, the Wilhelms and Kalmons, in a roundabout way. After Theno made pro in the professional golf circuit in the mid-1960s—and in just four short years instead of the typical five—it seemed as though his dream of making the U.S. Open someday was closer to becoming a reality. But it was not to be.

Several years into his golf career, Theno's daughter was diagnosed with terminal cancer. He had no health insurance and, unless he wanted to be strapped with medical debt for the foreseeable future, he had to get a good policy. And quick.

Dick Hunegs, the firm's senior partner and lead attorney for the Fitzgerald cases, had met Theno on the links. Theno had helped Hunegs and his partners brush up on their chip shots and putts. Hunegs enjoyed Theno's company. He also knew of his family situation. So, he offered the affable and talented young father a full-time position at his firm. If he was half as good with legal work as he was with his golf game, his firm would be getting a steal. In return for his labor, Dick Theno would be eligible for blanket health insurance that would cover his ailing daughter's health bills in full. "There was no decision to be made," Theno said. "I went to work for them."

Theno lost his daughter to brain cancer soon after. His loss—albeit now five decades removed—haunts him to his day. His gray, downcast face said as much. He tries to punch through the sadness, but a forced smile betrays the long-held melancholy.

Theno no doubt could have returned to his golf game. "If I had not gone to work for Dick [Hunegs], I'd still be playing professional golf today," he told me. But in his mind that is not the way the world works. You dance with the one who brought you. Theno was eternally grateful for Hunegs's generosity when he was at the lowest point of his life and would stick by his side for the next eight years, which included the time the firm was handling the two Fitzgerald cases.

George "Red" "Shorty" Burgner was the kind of witness that only comes around once in a lifetime. And he was a fun one, too. You know you have personality when you have two nicknames. "He was a rich character," Hunegs told me of his prized witness.[20]

Like Dick Theno, Burgner was a proud son of the Northwoods. This Superior native found himself on the Great Lakes at age fifteen. He only finished the tenth grade and was soon drafted into the U.S. Army. He saw action in the European theatre of World War II between 1942 and 1945 and fought on the front lines, where he rose to the rank of Sergeant in the 26th Infantry Division. He fought in France, pushed through into Germany, and liberated a concentration camp in Nazi-occupied Austria.[21] Upon his return home, Burgner went straight to the front office of Oglebay Norton Corporation to get his old job back. The shipping line gladly took him in, and he returned the favor with thirty years of dedicated service to the company. He sailed on twelve ships, from steamers to iron ore tankers. The Fitzgerald was his last assignment.[22]

Red was happy to spill the beans on Oglebay Norton Corporation. The Fitzgerald was not seaworthy and had been neglected for years. And the negligence went straight to the top. He had taken the president of Oglebay Norton Corporation, Rennie Thompson, on a tour of the ship itself in the winter of 1973–74 when the keel was crumbling apart and mounds of lake water dirt and sludge had accumulated in the ballast tank, rotting out the hull (see chapter 6).

Red was the key to their case. But he was less than perfect. And his adopted hometown of Chinero, Texas, was not a paragon of racial equality. White townsfolk threw around the N-word with abandon. Red always had a shotgun at the ready in case someone who didn't look right stepped foot on his property. His truck was well stocked with beer, and he could be seen around town with a can in one hand and the other hand on the steering wheel.

In the fall of 1977 when Theno visited Burgner, Theno could hear the sounds of sawing and nailing coming from the downtown area just a few blocks from Red's garage. Local authorities were building a scaffold at the town square that could be used if the verdict of a pending murder trial turned out the way they thought it would. Red was well acquainted with old-time justice and the wild west, and he had a nose for it. That instinct would soon come in handy.

Two representatives of Oglebay Norton Corporation and Northwestern Mutual—Thomas Murphy, attorney for Oglebay Norton Corporation and Northwestern Mutual, and Robert Kratzert, vessel personnel manager for Oglebay Norton Corporation—had called Burgner up shortly after the ship's sinking and strongly discouraged him from attending the Coast Guard's hearings in the inquiry of the ship's sinking. Incidentally, on the other side of the buttered bread, Captain Jim Wilson, a Coast Guard investigator, told Burgner that "his testimony wasn't needed."[23] "We don't want you up here. We don't even want your name mentioned. . . . It's a good place for you down in Texas," Kratzert and Murphy said, according to Burgner's deposition of December 13, 1977. He did not attend and was not a witness at the Coast Guard hearing, whose report was published on July 26, 1977.

According to the transcript of the deposition, the representatives from Oglebay Norton Corporation and Northwestern Mutual had told him to keep his lips sealed if any nosy attorneys—or private investigators—came snooping around for testimony that could implicate their clients. "Don't talk to the papers, don't talk to nobody about it. If anybody asks you anything about it, don't say anything. And if you get any contacts from anybody, don't say anything to them until you contact us first," said Kratzert. At first, he abided by their wishes because he wanted the companies to do the right thing and take care of the families. "All I was concerned about at the time was they were gonna help the people that the families [*sic*] and that stuff," Burgner said. "And I told them that at that time, 'You take care of the families. Treat them right,' and I says, 'I won't say nothing to anybody. I'll keep my mouth shut.'"[24]

Murphy and Kratzert told him not to worry. But they quickly reneged on their promise. Red had gotten a telephone call from Allen Kalmon's widow, Immy Kalmon, who said the company was not returning phone calls. Red and Allen were tight and had sailed together for years.

Kalmon was from Ashland, and his widow was introduced to Theno via a high school friend of Theno's, Pat Simmons, daughter of Wheelsman John Simmons. (Simmons was likely in the pilothouse when the ship went down.) Theno contacted the Simmons family before they spoke with Oglebay Norton Corporation. The two families knew each other from the same social circles. Theno advised them not to settle with the insurance company and that if they needed legal help, he was there for them. Nonetheless, the Simmons family settled with Oglebay Norton Corporation directly for $75,000, a fraction of the amount some other families would receive from their settlements.

While it was too late to help Pat's family, it was not too late to help out the Kalmon and Wilhelm families. "I can remember that clearly," Steve Eckman, former attorney of the DeParque firm who helped represent the Kalmon and Wilhelm, said of the conference room where he conducted the discovery deposition.[25] He told me this from his home about a half-hour drive west from his old law office in downtown Minneapolis. The elderly, though young at heart, Eckman was laid up on a sofa with a heavily bandaged right ankle from a tumble down the stairs at his secluded rural home just outside Independence, Minnesota.

Eckman grabbed the legal pad from my hands and began to sketch a mockup of the conference room for the deposition. "I can remember who was sitting where and where I was sitting," he told me as he began to draw. It was like any old conference room, Eckman told me. He drew what appeared to be a rectangle of sorts along the wall for bookcases, a wedge for doors at either end of the big rectangular room, and the coup de grâce, the head of an oblong table where prized witness Red Burgner would sit.

But Burgner was not represented by a stick figure or a hastily drawn caricature. Eckman made sure to carefully outline arms, legs, and a torso; Eckman appeared to me respectful and reverent. Forty-seven years later the memory was just as potent as the experience itself was on that early winter afternoon of 1977.

The discovery deposition was the Wilhelm and Kalmon families' response to Oglebay Norton Corporation and Northwestern Mutual's Limitation of Action filed in the United States District Court for the Northern District of Ohio, Eastern Division. A Limitation of Action is a company's typical response in industrial accidents like ship sinkings. It is an effort to cap the company's liability at the value of the ship in its current state, which in this case was a wreck at the bottom of Lake Superior.

The case was filed in federal court in Ohio as opposed to Saint Paul or the western district of Wisconsin because it was joined with other Fitzgerald cases that had already been filed in that particular court. Because it was a maritime case, it could be heard in any number of venues, according to Eckman. While he doesn't have a clear memory on the rationale for the choice of venue, this explanation made the most sense to him in retrospect.

Red was taking a chance and putting his life on the line to help his friend's widow and family. Theno said that Burgner thought Oglebay Norton Corporation and Northwestern Mutual's threats were serious, and they could very well act on them.

It made sense. There was a lot at stake; it was a high-profile case. By 1977, the *Edmund Fitzgerald* was a household name thanks to Canadian folk singer Gordon Lightfoot's popular song "The Wreck of the Edmund Fitzgerald," which spent twenty-one weeks in 1976 on the U.S. Billboard charts.[26] It remains a staple track on classic rock radio stations.

Red feared for his life. And Theno believed him right away. Theno easily convinced his boss, Dick Hunegs to get him to a deposition. "The decision was to get Burgner into Minneapolis–St. Paul, as quickly as possible," Theno told me.

On the trip to Minneapolis, Theno took extra caution to make sure he'd arrive in the Twin Cities safely, which included putting him up in back road motels under aliases. "Theno got him a route and gave him reservations along the way that would not be in his name because he was convinced they would chase him down if they knew he was headed up that way," Eckman told me.

Neither Theno or Hunegs or any other attorney had experienced a situation in which a witness's life was threatened. "I never heard that in any other cases," Hunegs told me from the comfort of his St. Louis Park, Minnesota, living room forty-five years later. "I have heard and have known about destruction of evidence, but I can't say in any case that I had other than [the Fitzgerald cases] that someone feared for their life." Dick practiced law for sixty-two years.[27]

It was well known that Red knew the Fitzgerald like the back of his hand. He sailed with her for nine years (1966–75). He could practically guide himself around the ballast hold blindfolded. No one else alive had sailed on the Fitzgerald so recently *and* knew its faults and strengths inside and out. If Red talked, it would be game over for Oglebay Norton Corporation and Northwestern Mutual.

Red's testimony in the discovery deposition was more than enough for the attorneys of Oglebay Norton Corporation and Northwestern Mutual to hear. At the conclusion of attorney Stephen Eckman's time with the witness—Eckman was the attorney from the De Parque firm that conducted the deposition—Murphy "surprised" Eckman by declining to ask any questions. "My assumption was that they were going to obviously cross examine him and try to diminish the impact of what he was saying," Eckman told me. "And when I said 'no further questions,' and I turned to them and said, 'Your witness,' they just looked at each other. They folded their papers; said 'no questions' and they stood up and walked out. That was it and I never dealt with them again."

Burgner truly believed he was putting his life at risk, and both Dick Theno and Dick Hunegs agreed. In a Minneapolis *Star Tribune* story on November 10, 1997, twenty years after the deposition, Kratzert, the retired personnel employee of Oglebay Norton Corporation, offered a dismissive, non-answer answer about threatening Burgner against testifying at the hearing or speaking to legal representatives for the families. "I don't remember that. That's pretty vague. I don't recall that," Kratzert said.[28] Interfering in a U.S. government investigation and threatening witnesses is a serious charge. It would follow that a strong denial on the part of the accused would be in order. But it was not.

Moreover, Burgner had nothing to gain. All settlement money went to the families—along with the contingent fee to the attorneys—and not him. Eckman put on the record at the deposition that his firm did not pay Burgner anything beyond travel expenses from Texas to Minnesota and back. And finally, recall that when given the opportunity, Murphy did not ask Burgner any questions during the deposition. If he *knew* Burgner was not a credible witness, Murphy could have shown that with a simple line of questioning. Surely an able corporate attorney like Murphy for one of the country's top shipping companies could handle that task.

There was one potential conflict of interest. According to Eckman, he recalled Burgner had a pending workers' compensation case before Oglebay Norton Corporation from the medical condition that kept him off the Fitzgerald at the last moment. However, by going public with incriminating knowledge of the condition of the Fitzgerald and, to say the least, unprofessional behavior of the company's representative, eliminated any leverage he had over the company to get a good settlement on his claim. "His only value to them was if they could get him taken care of either by settlement or harm's way before his testimony was preserved like this," Eckman told me. "Once this was in the books, his story was out. Any leverage that he had at that point was gone."

Within a couple of months following the deposition, the companies settled. Imny Wilhelm settled for $187,500 (nearly $1 million in 2023). Kalmon's settlement was over that amount.[29] "I feel much more—much more suspicious that it was not an accident. You know, there would be very few places where those legal secrets would lie. And they would lie among the lawyers' files," Dick Hunegs, lead attorney for the Wilhelm and Kalmon families told me. In front of the bay windows of his living room inside his suburban Minneapolis ranch home, with family photos on top of a baby grand piano in one corner and keepsakes from a life well lived inside a gleaming glass case in the other, Dick stared at me in stunned silence after I told him that Toby Marcovich's basement had flooded shortly after the Fitzgerald cases were closed.[30] Incredibly, Hunegs's law office basement had flooded, too.

For the better part of a storied sixty-two-year legal career, during which he left an indelible print on consumer and tort law that will ensure justice for victims for a long time to come, Dick Hunegs, the Minneapolis version of Toby Marcovich, was haunted by that lingering event in his life. All of the firm's case files were ruined, including the files of the Fitzgerald cases. "We had experts in to find out what happened to the files. But it was just a bizarre, bizarre happenstance," he told me. In the comfort of his home as a late-morning sun poured into his living room, Dick looked as though a weight had been lifted off his shoulders but another had been put in its place. The mystery was not solved.

Neither Dick nor Toby—who had not suspected that the cause of his basement flooding was anything other than an unfortunate accident—would likely ever know the true nature of the flooding nor the reasons behind it. Dick was not going to get closure, just as the Fitzgerald families would never learn the precise reasons for the ship's sinking or the details of the final moments of the lives of their loved ones. Memorials and remembrances offer comfort, but they never offer closure. And the families likely will never get it. "There's a memorial service. There's no funeral," Deborah Champeau, eldest daughter of Third Engineer Oliver Champeau, told me. "I look at funerals as, 'You're lucky. You can see him one last time.' It's really hard when you don't."[31] Dick Hunegs was okay with how things ended up for him. Or at least that was the face he put on for his family. "You made my dad's day," his youngest son told me later that evening.

CHAPTER 6

A Bone to Be Chewed

IN 2010 GORDON LIGHTFOOT CHANGED ONE LINE IN HIS SONG, THE "WRECK OF THE Edmund Fitzgerald." Instead of "At 7 p.m. the main hatchway caved in," he would now sing, "At 7 p.m. *it grew dark, it was then.*" It seemed innocuous. What's one line? Especially in pop songs where half the lyrics can be notoriously undifferentiable ("Benny and the Jets," anyone?).

Lightfoot simply exercised his right to change his mind because he had new information. And, unlike Elton John, Lightfoot felt it was important to be crystal clear about his subject. "He was pretty impressed by what he saw in the film, new evidence that unsecured hatch covers didn't cause the ship to sink," a Lightfoot spokesperson said when questioned about the crooner's decision to change his lyrics.[1]

The spokesperson was referring to a 2009 television episode of the documentary series *Dive Detectives*. It only ran for a season, but the material they were able to dig up to shed light on new theories of the ship's sinking was worth the production expense for the full series. In the episode, the host presents strong evidence that a rogue wave approximately fifty feet high did the Fitz in. It was considered a fringe theory until recently, when, according to a leading *Edmund Fitzgerald* historian and author, scientists conducting research on Lake Erie used reverse radar on the lake bottom to detect waves out of sequence.

Incredibly, the initial U.S. Coast Guard report that was released on July 26, 1977, still stands as the U.S. government's official finding, forty-six years running.[2] The culprit: damaged hatches, as formerly memorialized in Lightfoot's ballad. The thought at the time went that the ship took on too much water in the forward hold because the hatches weren't covered. On its surface this seems ridiculous. Why not consider new information and explore new theories? Much has changed since the Coast Guard released the report: multiple dives to the wreck, reams of photographs and videos, new technology and modeling, as well as the Internet to gather and meld the minds of the Fitzgerald diaspora. The Polish authorities ultimately changed their minds about Copernicus's wild ideas. Why won't the U.S. government consider new theories?

It's not so much that the U.S. Coast Guard truly believes the main cause was the hatches giving way. Rather, they established an "institutional consensus" that they want to protect. By that I mean an arrangement—spoken or unspoken—where affected parties accept their role and responsibility and deal with the ramifications, whether they like it or not. It's for the greater good. A harmless conspiracy of sorts.

For the subjects of the investigation, there is little to be gained by reopening the investigation. For the U.S. Coast Guard, it could mean a sullied reputation or a reminder that it botched the rescue attempts. For Northwestern Mutual, it could be a bad press run that could seep into their customer base and undermine confidence in the company. Northwestern Mutual, the largest insurer in the country, has a brand to protect and a customer base to keep happy. Despite all affected entities, just one organization, the Lake Carriers Association, challenged the Coast Guard's report with its own study and findings.

One line that remained in Lightfoot's masterpiece, and should always remain, is "that good ship and crew was a bone to be chewed," though not for its intended meaning. It wasn't a "bone to be chewed" because of the rough winter waters that it would experience but because of blatant corporate malfeasance and flawed and ineffective governmental oversight. The Fitzgerald was a bone to be chewed by the very same economic and governmental forces that brought down an entire economy.

On June 8, 1958, the *Edmund Fitzgerald* splashed into the Detroit River from its dry dock at the Great Lakes Engineering Works (GLEW), where masons, carpenters, welders, and pipe fitters had poured their hearts and souls into its construction over the past thirteen months. She was so massive it took thirty-six minutes for the workers to just untie the keel straps. When the ship slid into the waters, its splashdown drenched the assembled crowd of fifteen thousand and nearly knocked the pier over. One spectator collapsed from a heart attack and died. It was quite an entrance.[3] A moment earlier, the wife of the ship's namesake, Mrs. Elizabeth Fitzgerald, had the honor of christening the ship with the traditional smashing of a bottle of champagne against the keel. It was an inauspicious start. It took Mrs. Fitzgerald three whacks before the bottle finally broke.

Fifty-four years earlier, in 1904, GLEW birthed its first boat, the RW *England*. Company founder Anthony Pessano wasn't one for a cautious entrance. Instead of gently sliding into the market by building a modest vessel, he went the distance and built a full-size freighter. Then again, Pessano and GLEW were the scrappy competitors, never afraid to go toe-to-toe with bigger competition. In 1958 they continued the tradition of unbridled ambition and tenacity by building the largest freighter that ever plied the Great Lakes.[4]

On the platform of the pier, the workers, designers, architects, crew members, local dignitaries and politicians grinned for the cameras. The crew members would be giddy in anticipation of being the first to steer the Fitzgerald when she set off the next day on her first voyage.

If one listened closely, one may have heard shop talk among the politicians. Republican Michigan U.S. Senator Charles Potter was in reelection mode and in the fight of his political life. He would fall to Philip Hart in a Democratic wave later that fall.[5] On the other hand, Democratic U.S. Senator Patrick McNamara would have been a pig in mud. He was a pipe fitter by trade and president of his local union for twenty years.[6] He could well appreciate the accomplishment of his union brothers and sisters.

But the architects and designers of the Fitzgerald may have been more sanguine. Save for the owners of the ship, Northwestern Mutual, they were the only ones who knew that the ship's builders had cut corners in its design. In subsequent years, according to Third Engineer Oliver "Buck" Champeau's daughter Deborah Champeau, designers of the Fitzgerald would approach her with apologies for its shoddy design.

"One engineer [designer] apologized that they only had this certain kind of steel that they could use on the ship because that's what they were given and they were on a budget," Champeau told me.[7] According to Red Burgner's deposition from December 13, 1977, for Stephen Eckman, attorney for the families of Allen Kalmon and Blaine Wilhelm, some of the Fitzgerald engineers knew "she wasn't a solid build."[8] (He did not name Oliver Champeau in the deposition.) Also, in the Burgner deposition, he cites a conversation with a foreman at the time of the Fitzgerald construction that the ship "'was built under production welds.'" A production weld is of low quality in that it is done fast and without any quality controls.[9] Additionally, according to Burgner in his deposition, the ship's former chief engineer, Wendel Freeman, who was at the construction yard "when [the ship] was started and finished, and he sailed on her . . . wasn't happy about the work that was done on it." According to Burgner, it was not a "solid weld" because it was "tack welded." Burgner also suggested the crew did not just cut corners in the welding but surreptitiously avoided Coast Guard scrutiny by doing the welding at night when the Coast Guard was not on hand. "Coast Guard would go away at night, tack weld it. Paint it. Cover it up, and they'd pass it the next day," Burgner said.[10]

A conspiracy that has been batted around but can be easily dismissed is that the ship was extended in its mid-section after it was built. It was not uncommon for companies to lengthen ships for the purpose of hauling larger loads. Of course, that is easily ascertained, and there was never any evidence or record of the Fitzgerald being lengthened.

Sault Ste Marie native and twenty-five-year veteran of the Nuclear Navy Mike TenEyck wrote an extensive white paper on the sinking of the Fitzgerald and explained to me how the ship could exhibit characteristics of having been extended but without actually being lengthened.[11] It made so much sense. Why hadn't those theorists thought of it?

"They took a 600-foot design and stretched it out," TenEyck told me. "They took the 600-foot design and extrapolated it to 729 feet."[12] By that, designers simply "made the ship longer and did not address the effects of the extra length."[13] This is the argument that marine manager Ray Ramsey made in his book *S.S. Edmund Fitzgerald: Requiem for the Toledo Express.*

That wasn't all of it. The wrong design cascaded into further problems. Because the boat had more capacity, it needed more power. It needed a bigger engine, according to TenEyck. But it would be akin to putting bigger pedals on a longer tricycle. It wouldn't match up. While the rider would have more leverage and torque, once the pedal did a full revolution it would strike the pavement, and the rider would topple over.

It's too bad these problems didn't come to light during the Coast Guard's investigation. They should have. There were no technological or physical impediments to gathering the information. All it would have taken was subpoenaing GLEW's blueprints. "You could argue that the Coast Guard or the government did some selective reporting," TenEyck told me. It was important to note who was interviewed as well as who was not, he pointed out. The lead investigator for one of the law firms representing two families in the follow-on legal actions identified a key witness who was directed by the Fitzgerald's owner and operator not to testify at the Coast Guard hearing.

And it gets worse. Corporate America would not learn its lesson and because of that, the same mistake would be repeated forty-three years later when two Boeing 737 MAXes fell out of the sky in 2019 and 346 souls perished. Boeing did something close to what the Fitzgerald designers had done, TenEyck pointed out. "Boeing took a regular 737 design and extended it to a 777 size to hold more passengers," he told me. The two planes were done in by the size of the engine that Boeing built and put into the modified aircraft. "They [the planes] were physically bigger. They [Boeing] had to go into the two engines on the 737 and make them bigger," TenEyck told me. "They had to move them slightly and also raise them because they would hit the bottom when they landed, hit the ground. By doing so, the modifications would move the center of gravity and affect the aerodynamics."

Like a bad Rube Goldberg contraption, one thing led to another and to another, and in the end it fell apart. TenEyck explained, "What they did instead of fix[ing] the fundamental problem so it would fly like a plane, they had this electronic monitoring system to make adjustments. It gave signals to the pilot, but it also caused it to accelerate as it stalled." The president of the Association of Flight Attendants union, Sara Nelson, commenting on the two MAX 737 flights, diagnosed the problem. She identified an intricate conspiracy of lax oversight, laser-like focus on profits, and incompetent political leadership, going to the highest levels of government and to Boeing, the largest defense contractor in the country.

"The 737 MAX was rushed to market without proper review, enabled by the out-of-control ODA [Organizational Designation Authorization, a Federal Aviation Association program] system that was put in the hands of [Boeing] to maintain," Nelson said. "Outsourcing regulatory oversight is a slippery slope as corporate efforts to maximize profits come at the highest cost. The crisis was exacerbated when the Trump administration kowtowed to Boeing and dragged its heels—damaging public confidence at home and the reputation of American aviation abroad."[14]

Also, it would seem that transport was one of few industries where engineers would extrapolate the size-design. "Civil engineers know never to take on design for a bridge and extrapolate it to a larger size," TenEcyk told me. "And when you do it, they end up collapsing; you just don't extrapolate things."

Ships like the Fitzgerald are meant to be flexible. If they weren't, they would snap apart. It's the same concept with skyscrapers. There has to be some give when wind kicks up, or in the case of buildings along a fault line, when tremors shake the foundation.

From the aft (i.e., stern), the ship would look like a diving board springing up and down. It is designed to react in that way. "If a ship is completely rigid in a seaway, you [would] break in two," Cooper said.[15] A correctly built ship would move up and down, not right to left as the Fitz would sometimes do. That would move the center of gravity off-balance and cause undue stress to the ship, compromising its structural integrity.

TenEyck recalled the description that Captain McSorley used to describe the movement. "McSorley used the words, he 'did not like the ship's wiggle." McSorley, as TenEyck pointedly recalled to me, likened the movement to a "girl's wiggle." In the words that only a serious engineer would use, TenEyck told me, "She's torsional around her hips. That's a wiggle. Why do people say that? Because that is her structural build." Had the sailors not been to sea so long and away from port, perhaps they would never have noted the difference in that way and thus identified a central defect of the ship.

Red Burgner also observed movement that was equally problematic. While the ship was designed to bend like a diving board, it exhibited two concerning movements. First, when the boat would hit a wave, the bow would thrust up but stay in that position. The effect was not unlike a motorcyclist popping their front wheel and holding it in air, or a speedboat cutting through waves with its bow inclined. Second, Burgner described a similar phenomenon to TenEyck's wiggle—a side-to-side (port to starboard) movement. "The Fitzgerald it seemed like the bow would just never get back there again [level on the water] and another one [wave] would hit it; she'd go back over again. Or it could be port or starboard, whichever way she was hitting . . . to all of us on the ship at the time it wasn't ordinary."

And just as McSorley did not like "wiggle," he also did not like the movement that Burgner described. "[McSorley] did not like the action of the ship."[16] Both types of movement caused undue and untold stress on the ship's structure that no doubt contributed to its ultimate demise.

Bob Hom, a mentee of the surly McSorley, recalled a time when he was a deckhand on the SS *Armco*, another freighter in the Oglebay Norton Corporation fleet that McSorley captained. As they passed the Fitzgerald, Hom said, "McSorley looked over at the Fitzgerald and said [to me], 'I'd hate to be on that boat in a big storm. It's all worn out from years overloading it.'"[17]

Red Burgner was the sailor who knew the Fitzgerald better than anyone else and lived to tell his side of the story. He had sailed the Fitzgerald for several years and kept ship, staying over in the ship and tending to it in the off-season, including the winter of 1973. He was on the Fitzgerald until early fall of 1975, when his doctor ordered him off the vessel because of bone spurs. He had collected plenty of vacation time, and it was an easy decision to ride out the last weeks of the season in his home in east Texas.

In great detail and at great risk to his life, Burgner gave a two-and-a-half-hour deposition on December 13, 1977, in Saint Paul, Minnesota, to Stephen Eckman, the attorney representing the families of Allen Kalmon and Blaine Wilhelm. Burgner was convinced his life was in jeopardy because, according to Burgner, representatives of both Oglebay Norton Corporation and Northwestern Mutual leaned hard on him not to testify at the Coast Guard hearing (he did not testify) and made clear on multiple occasions that they did not want him talking to any attorneys of the Fitzgerald families. Burgner fervently believed that the Fitzgerald was totally unfit to sail, especially in the worst weather month of November. He saw firsthand the condition of the bottom of the keel.

For years, the Fitzgerald's owners maintained her on the cheap. Despite regular inspections by the U.S. Coast Guard, the Fitzgerald's condition was overlooked and the ship approved for sailing every time. The keel, the backbone of the ship, was especially problematic. If the keel failed, the ship would fall apart, just as any animal would collapse if it had a cracked vertebra. You could say that the Fitzgerald was riddled with cracked vertebrae up and down its keel.[18]

As recently as 1973, two years before the ship sank, a loose keel was discovered by Red Mead, a foreman for Fraser Shipyards in Superior, Wisconsin, where the Fitzgerald wintered in port that off-season. The welds had "broken loose" entirely.

"Red Mead shoved the crowbar right under it and what he pushed out of there was welding rods and everything else when the ship was built," Burgner testified in his deposition.[19]

It wasn't too much to fix the keel because they were already opening the Fitzgerald up for other repairs. The Fitz had been blackened and blued. They had to replace "the stern, section top and gunwale, which is the rim of the hull,"[20] according to Burgner. Former Fitzgerald captain Peter Pulcer had backed the Fitz into the steamer *Sylvania* and caused severe damage to its structure. They also used this opportunity to install a new bow thruster, which is used to propel the ship in a lateral, side-to-side direction. Whether it was installed for the purpose of correcting the ship's "wiggle" is not clear.

To fix the weld, they had to work in an area that was no taller than six feet and clear out mud that was in places up to four feet deep. For many industries this working condition would raise a red flag and call the attention of OSHA. Not in the shipping industry. Safety and workplace conditions were regulated solely by the Coast Guard, whose enforcement record was spotty at best.[21] Burgner estimated they cleared between a twenty- and two hundred-foot length of keel of mud to get at it the broken weld. To repair it, the crew simply welded two pieces of metal together, known as a tack weld. According to Burgner, such a fix was not a serious repair. "It's not solid the way it should be. Because all your [*sic*] welding is a welding rod across."[22] It was a patchwork job, nothing remotely resembling a serious structural correction.

There's more. The loose keel wasn't entirely fixed. A portion of the keel still needed to be repaired, but the Coast Guard gave them permission to delay it until the following winter. For now, according to Red Burgner, the U.S. Coast Guard deemed the Fitz "seaworthy enough to sail."[23] The Fitzgerald was good to go for the 1973 season.

Perhaps the most troubling part of the finding was not just that it was sailing in that condition but that the company knew it was unseaworthy, and they had every opportunity to fix the structural defect the right way by rebuilding the entire keel and not just patching it over, but they *chose* not to. And that knowledge went straight to the top—Captain McSorley and Oglebay Norton Corporation executive Renold Thompson.[24] Besides, the Coast Guard wasn't a force to reckon with, and Oglebay Norton Corporation often had their way with them, like Red Burgner, who reportedly prepared a surf and turf dinner to get cleared for an entire season of runs (see below).

According to the deposition of Red Burgner, Ransom "Ray" Cundy, William Spengler,[25] and Eugene "Red" O'Brien approached McSorley in 1975 about the condition of the keel.[26] This was the exchange in the mess room that Burgner recalled two years later:

> **CUNDY:** Captain, the keel's loose on this son of a bitch again.
>
> **MCSORLEY:** I don't give a fuck. All this son of a bitch has got to do is stay together one more year. After that I don't give a shit what happens to it.

In the 1995 documentary *Expedition '94: To the Fitzgerald*, Red Burgner recounted a conversation he had with Thompson and Red Mead of Fraser Shipyards.

> "He [Thompson] says 'What's that hole in the back open for?'
> [Red Mead] said, 'They're putting the weld in the keel.'
> 'What do you mean?' Thompson said.
> [Red Mead] says 'The keel is loose.'
> 'You're kidding?!' [Thompson said.]
> 'No,' [Red Mead] says, 'You want to see it?'
> 'Yeah,' [Thompson said].
> [Renold] Thompson. We all walked in there. Just the three of us. There's a ladder going down. Water bottom there is six feet or better. He walked on the bottom walked around a little bit and [third man] says, 'Here. I'll show you.' He took a number 16 scoop shovel, and shoved it underneath the keel and pushed out welding rods."[27]

Thompson and Oglebay Norton Corporation were caught red-handed.

In this context, Sara Nelson's comment regarding Boeing, "put in the hands of [the corporation] to maintain," sounds eerily familiar. It sounds like the arrangement in the banking and finance industry, where outside, private entities—in this case credit agencies like Moody's or Standard and Poor's—funded by the companies that they regulate, determine the soundness and quality of the product. It was this arrangement that led to the financial meltdown of Wall Street in 2008 and, subsequently, the Great Recession. Moody's gave sub-prime mortgage securities Aaa bond ratings even though the mortgages were risky bets. It was a broken business model that put the entire economy at risk.

Like any private enterprise, credit rating agencies exist to turn a profit. And because it is a competitive market with several credit agencies, including Moody's, Standard and Poor's, and Fitch, businesses would shop around for the best deal. "The market competition between the ratings agencies . . . actually made them worse," wrote Bethany McLean and Joe Nocera in *All the Devils are Here: The Hidden History of the Financial Crisis*.[28] Moody's paid a settlement claim of $864 million to the U.S. government and states; Standard and Poor's, $1.375 billion.[29] A drop in the bucket compared to the nearly $8 trillion in stock market value wiped out between 2007 and 2009.[30]

Lax oversight from the credit agencies was nothing new. One reason New York City fell into a major fiscal crisis in the mid-1970s was because credit agencies like Moody's continued to give the city high bond ratings, even raising them while the city fell deeper into debt and noting the city's "amazing resiliency." Besides, the "city's problems were nothing new," credit agencies assured investors, according to author Kim Phillips-Fein's account of the crisis. "For half a century now, it has been widely known that New York City has a revenue problem," Moody's stated on the occasion of yet another bond issuance.[31] The American Bureau of Shipping (ABS), a nongovernmental agency tasked with evaluating the seaworthiness of vessels in the shipping industry wasn't much better. Ships are reevaluated for seaworthiness when the vessel is extended or modified, or the owner contends that it is capable of carrying additional cargo. Like credit agencies, ABS is funded by private industry, in this case the shipping companies. This arrangement calls into question the legitimacy of the evaluation, just as credit agencies were questioned in the throes of the financial crisis in 2008.

There could have been several instances when ship companies leaned on ABS or the Coast Guard to give them the benefit of the doubt and their seal of approval. In one story that Burgner shared in *Expedition '94*, it sounded like a backroom deal at a political convention.

> "And George Holl [Fitzgerald chief engineer] said to me, 'Red we ain't going to get any inspection sticker for this sucker this spring.'"
>
> "Why?"
>
> "Too damn many restrictions."
>
> "What can I do?"

"I don't know, maybe you should—you know call the coast guard guys; you know, call ABS and everything like that."

"Well how about a big dinner?"

"You know this is costing the company money. It's gonna cost you" [Burgner told the Coast Guard and ABS employees].

"Really? What's this all about Red?"

"Well, George Holl is wondering if he is gonna get his inspection sticker."

"No problem, no problem."[32]

According to *Expedition '94* director and narrator Ric Mixter, the Fitzgerald would get its inspection sticker and would break a milestone hauling record by clearing 30,269 tons of taconite, "more than any other ship in history," in 1968. According to Burgner, the ship, under Captain Peter Pulcer, would go on to break its own record six more times. "She was tired. You know, they pushed the hell out of her," Burgner said.[33] Sometimes it was loaded up so much that the deck of the ship rode just ten feet above the water line. If Chicago Bulls fans were right and Michael Jordan walked on water, he could have dunked on the Fitzgerald.

This wasn't the only time that the company got their way. Sometimes when ships came through the locks and were riding above their Plimsol mark (i.e., carrying more weight than they were allowed), a crew member would toss a little gift to the lock tender to get cleared.

Tim Nixon, a former sailor on the Great Lakes and Green Bay, Wisconsin, attorney, recounted a not so irregular practice when ships came through the Sault locks slightly overloaded. According to Nixon, there was "not a lot of abuse" with this practice, but "there were times that you might accidentally have put in too much cargo in a certain place."[34] Nixon continued: "So, you come into the Soo Locks, you had the guy comes out to read your marks [on the stern]." (The Plimsol mark is on the side of the ship and unobservable when the ship is in the locks and nearly flush against the wall. In that case, the lock tender or deckhand extrapolates where the draft would be on the Plimsol mark based on the water level at the stern of the boat.) "And I have a three-pound bag of coffee in my hand," Nixon pauses to try and hold in his laughter. He sounds like a raconteur at a bar at 2:30 a.m. "And I go, 'What do you read those marks as' and I would hold up the bag of coffee. And he goes, 'What do you want him to be?' And I'd say '26 [feet], 3 [inches].' 'Yep. 26'3" on the spot.' 'I toss him the bag of coffee.'"

The Sault was the only stop along the route where someone besides a crew member checked and reported the ship's draft relative to the Plimsol mark. When the ship set off, they were only required to self-report the information. "When you left the dock, you would tell how deep you were," Nixon told me. "But nobody else looked at it but one of the ship's officers. And you would log it and report it." Every once in a while, a surveyor would board the ship and measure.

The incentive to load up as much taconite as possible fell not to the crew members but to the captain and by extension the company. Captains were the only ones on a ship that reported to the company. Everyone else was a member of the Lake Seamen's Union, an arrangement akin to trade crafts where construction companies get their labor pool from local union halls. Captains might get a little extra pay, but the companies would be the real beneficiaries.

The Fitzgerald was carrying thirty thousand tons of taconite in 1968 instead of its usual twenty-six thousand tons. This translated into a 15 percent increase in cargo hauls. That meant that Oglebay Norton Corporation got an extra free trip every five or six runs. Not bad. This would only compound and redound to the benefit of the company in subsequent years when loads were increased.

Regardless of where the incentive lay, it was the captain and crew who were taking the ultimate risk. Blame should fall on Oglebay Norton Corporation, its executive officers and its shareholders. They made the big bucks without the risk—not the workers.

"Some thought if [the Burgner deposition] were true it would be scandalous for the industry and most serious regarding the punitive penalty for neglect to make proper and timely repair to a long time known impaired hull resulting in the death of the entire crew of 29 crewmen."[35]

So wrote Richard Orgel, a long-time former Fitzgerald sailor. It would have made good copy for a *New York Times* investigative series or a good closing argument for a trial attorney suing a negligent company. Instead, it was part of a cover letter written to and filed away at the Great Lakes Collection of the Great Lakes at Bowling Green State University in Ohio. Lost in the files of the archives for the past eight years—the entry was incorrectly entered into the database by inverting the "g" and "n" in Burgner[36]—Orgel cried out to researchers, investigators, and the authorities about the crimes committed against the Fitzgerald and her crew. Orgel passed away in 2021.

To Orgel, Burgner's deposition was checkmate. Eyewitness testimony and a thorough timeline and catalog of the offenses, all under oath. And when the opposing counsel was allowed to question the subject, they demurred. "They just looked at each other, folded their papers, said no questions, stood up and walked out. That was it and I personally never dealt with them again," Eckman told me, reflecting on the deposition he conducted nearly forty-five years ago. The Burgner deposition would remain lost in library archives or hidden away in old office files in storage rooms. That is, until now.

CHAPTER 7

A Red Sea

I WAS BORN IN THE STRANGEST DECADE IN AMERICAN HISTORY: THE 1970S. I'M GLAD I have few memories of this time. I'm certain most Americans who lived at the dawn of the Age of Aquarius would agree.

Cults were in fashion and pushing some pretty wild ideas. There were pyramids with secrets and special powers, spiritual encounters with Mescalito—a "spirit" in peyote—and self-improvement methods that included auditing human bodies and discovering Thetans. The ubiquitous phrase "drink the Kool aid" was coined in the wake of the Jonestown, Guyana, cult mass suicide in 1978.

The American catalog of music resembled the product of over-consumption of cocaine more than anything else. The decade was bookended by two bizarre and forgettable music genres: progressive rock and disco. "Convoy," an ode to the citizens band radio in semitrucks, would spend six weeks at the top of the country music charts and end up selling twenty million copies. Even Betty Ford got in on the action with her own CB handle, "First Momma." Rick Dees's "Disco Duck" and "His Cast of Idiots" topped the charts. Listeners were put out of their misery when it was knocked off the number two spot by the hit folk ballad "The Wreck of the Edmund Fitzgerald" on November 20, 1976.[1]

The Beatles had disbanded. Elvis Presley was in a downward spiral and would leave us amid a haze of drugs, booze, and peanut-butter-and-banana waffles. The tenor of moviemaking had become dark by the '70s when compared to previous decades. The subject material was relatively sinister (*The Exorcist*), discomforting (*Deliverance*) and disturbing (*A Clockwork Orange*).

Americans were obsessed with conspiracies, notably the Kennedy assassination. The long-suppressed Zapruder film was aired for the first time in 1975. The U.S. House Select Committee on Assassinations was promptly formed in response to a public outcry and soon after reported out that President John F. Kennedy was "probably assassinated as a result of a conspiracy."[2] In 1976, disgruntled ex-NASA contractor Bill Kaysing invented the conspiracy that we never actually went to the moon in 1969. The conspiracy gained traction yielding an array of theories; one argued that it was all staged on a movie set directed by Stanley Kubrick, whose film *2001: A Space Odyssey* featured scenes that resembled those of the official NASA photos.[3]

Maybe it was the Vietnam War. Maybe it was the drugs. Maybe it was the oil embargoes. And the Watergate scandal certainly didn't help. But it was an odd time in American history to be alive. So, it's no surprise then that a country that was experiencing transformative social change was also struggling through a radical political metamorphosis from the left and right ends of the political spectrum.

President Lyndon Johnson's Great Society of the 1960s was left to rot along the wayside. National health insurance, the eradication of poverty, the elimination of hunger, and a guaranteed employment program would have to wait. *Roe v. Wade* (1973) would upend the country over the issue of abortion, a division that has yet to be bridged. In a relatively short period of time, the nation's political pendulum had swung hard toward conservatism.

The year that the *Edmund Fitzgerald* went down, 1975, marked not just the middle of the decade, but its political, economic, and social nadir. America had officially lost the Vietnam War with the ignominious fall of Saigon, which featured the iconic photograph of the desperate retreat of American CIA officers up the stairs, onto a roof of a hotel, and into an awaiting helicopter to whirl them away. New York City almost went bankrupt. Inflation rates reached double digits. Unemployment rates peaked at 9 percent. The president declared a recession. Two million workers went on strike. And the Middle East's Organization of Petroleum Exporting Counties (OPEC) raised oil prices by 10 percent. Nixon's resignation was barely a year old. The body politic was convulsing in shock.[4]

All of this was occurring as the country was preparing for its bicentennial anniversary, which would serve more as a reminder of how far we had fallen or how disappointed our Founding Fathers would have been with us than anything remotely resembling a celebration.

It was under this pall that former California governor Ronald Reagan launched a quixotic campaign for the presidency.[5] Movement conservatives revered Reagan like a biblical prophet, but he would need a divine miracle to strip the nomination away from a sitting president, let alone win a general election.

At first no one gave the former B-movie actor much of a chance, which was curious because his resume suggested he might make an ideal candidate. He served as chief executive of the country's largest state for two terms, boasted movie star looks, could deliver one-liners like Hollywood's best, was a former union president, and had a good following from years of being on nationally syndicated radio.

Reagan launched his campaign on November 20, 1975, ten days after the sinking of the Fitzgerald. Reagan's age and brand of politics were the only things his detractors could throw at him. And even that proved fleeting. Voters quickly looked past his sixty-four years. His stamina and endurance on the campaign trail and a punishing schedule proved the naysayers wrong. And soon voters would embrace his conservative political philosophy and warm to his messaging. His appeal would cross political lines. As former Alabama governor and segregationist George Wallace's candidacy in the Democratic primaries faltered, the Reagan camp took full advantage by courting Democrats without a political home in open primaries where they could cross over and vote in the Republican primary. In several contests, the conservative Democratic populist vote proved decisive, including North Carolina on March 23, Reagan's first victory. Later Reagan won Texas, where he trounced incumbent president Gerald Ford 66 percent to 33 percent, claiming all of the Lone Star State's ninety-six delegates.[6]

Not only did Reagan rack up delegates but he also captured a momentum that carried him successfully through even more contests. The two contenders duked it out all the way to Republican National Convention in Kansas City, Missouri, in August 1976. Accounts varied, but some had Reagan within fifteen delegates of Ford at one point. Later, Reagan claimed more delegates than Ford (1,140 to 1,135) but this was not accurate because the total of 2,275 exceeded the number of voting delegates.[7] It would be a real donnybrook when the campaigns met in Kansas City. And it would be close. Real close. The Reagan camp was so confident they'd pull it off, Reagan booster Joseph Coors had celebratory beers sent over to their headquarters at the Ramada Hotel in Kansas City on the eve of the convention.

It all came down to the notorious, mercurial Mississippi delegation. After weeks of playing hard to get and milking the president for all he was worth—taking rides on Air Force One, dining with the Queen of England, securing business loans, hanging out in the Oval Office—the duplicitous and transactional Mississippi state chair, Clarke Reed, threw all of his state's delegates behind President Ford. Mississippi held an exceptional waiver from the national party, since it was allowed to seat sixty delegates though each delegate had one-half of a vote, making its total thirty votes. This was intended to fold in new, younger leaders while placating the older generation.[8]

Reagan simply could not compete against the perks and overwhelming power wrought by the highest office of the land. This was as far as this insurgent, anti-elitist establishment candidate would be able to go in 1976.

Ford may have won the delegate count, but Reagan won the hearts and minds of the delegates. Ford's acceptance speech was an afterthought; no one remembers it. Reagan's carefully choreographed walk to the dais from his suite on the other side of the arena and perfectly delivered remarks—which included no endorsement of Ford—set the stage for his next campaign. His remarks were ad-libbed from an acceptance speech he would have given. By now, delivering prime-time stem-winders that marked major shifts in American political attitudes seemed routine for Reagan. He was well practiced.

In 1964, Reagan delivered what would be known as "The Speech" by the faithful followers of the American conservative movement. "A Time for Choosing"—as it was known then—was a thirty-minute televised speech on October 27, 1964, on behalf of Republican presidential nominee Barry Goldwater. It raised Goldwater $1 million, but it could not salvage Goldwater's wreck of a campaign. Goldwater went down to defeat the following week to President Johnson, 61 percent to 39 percent. But Reagan's prime-time address launched his own political career. He leapt over the usual trajectory of running for local or legislative office and ran and won the governorship of California two years later in 1966.[9]

The theme of "The Speech" presaged Reagan's message in Kansas City in 1976. Americans were called on not just to vote for Reagan's candidate but to take sides on a referendum on the philosophical future of the Republican Party. Would the party of Abraham Lincoln, Teddy Roosevelt and Dwight D. Eisenhower remain in the political center—a sensible variety of American conservatism that deferred to the free market while carving out just enough state support for the truly needy and also respecting personal liberty? Or would a new crop of candidates espousing a decidedly conservative line carry the mantle?

Reagan adroitly married American conservatism with populism in his speech, a foretaste of subsequent insurgent GOP candidates, including Pat Buchanan in 1992 and 1996, Ron Paul in 2008, and Donald Trump in 2016. "This is the issue of this election," Reagan said. "Whether we believe in our capacity for self-government or whether we abandon the American revolution and confess that a little intellectual elite in a far-distant capitol can plan our lives for us better than we can plan them ourselves."[10]

Populism was foundational to the address's message, but it was also part and parcel with an electoral strategy. Reagan knew that any successful candidacy would have to appeal to the grass roots and not the power brokers at national conventions or party bosses in smoke-filled back rooms. His anti-elitist message channeled that strategy well.

If "The Speech" raised the question of which of the competing Republican philosophies would lead the party into the next century, then Reagan's impromptu convention address in 1976 offered the answer. Conservatism was on the ascent. The 1976 GOP platform contained "bold, unmistakable colors with no pale pastel shades," according to Reagan—a not-so-subtle swipe at earlier, moderate to right-of-center platforms.[11] It was also a call to arms.

Reagan electrified the delegates and gave the crowd a sense of buyer's remorse. Nothing Ford could say would match, let alone beat, the address.

> This is our challenge, and this is why we're here in this hall tonight. Better than we've ever done before, we've got to quit talking to each other and about each other and go out and communicate to the world that we may be fewer in numbers than we've ever been, but we carry the message they're waiting for. We must go forth from here united, determined and what a great general said a few years ago is true: "There is no substitute for victory."[12]

Ford fell short in the general election by two points against former Georgia governor Jimmy Carter. While 1976 was a good cycle for Democrats, since they held on to both houses of Congress and picked up a governorship, Carter's victory was far from being a mandate. As a former southern governor, he came from the conservative wing of the Democratic Party, and it was hard to argue that progressive or liberal issues tipped the scales in his favor. Carter opposed abortion rights, unlike the party's 1972 nominee, George McGovern.[13] His foreign policy was decidedly hawkish and under his leadership America would commence a massive arms buildup that would carry into the next decade.

Carter did not salve the wounds of the body politic from punishing economic times. Working families saw nothing that benefited them: tax cuts for the rich, deregulation of industry, and failures to pass health insurance reform and labor rights legislation. Carter's choice to ignore these issues, particularly "his unwillingness to address the recession and anemic recovery" could have been predicted because he avoided the issues in his campaign. He believed he won on the basis of "morality and leadership" themes.[14] Major unions would abandon the Democratic standard bearer in 1980 and endorse Reagan. Bill Clinton would likewise abandon labor in the 1990s, but by then unions had little power to fight back and made little difference in election outcomes.

Soon it became clear as a brilliant California sky that Reagan had a wide berth to jump in and wrestle away the heart and soul of the American people from both the Democratic Party and the moderate wing of the Republican Party.

Reagan's first speech after Ford's loss in November of 1976 was an address to the Intercollegiate Studies Institute, a conservative network of college-age students, just five days before Carter's inauguration in January of 1977.[15] It was an ironic choice because Reagan was laying out an electoral vision grounded in winning over the traditional Democratic constituency of non-college-educated, blue-collar, white ethnic voters.

Perhaps it was a convenient audience, since the Republican National Committee was also in Washington, DC, that day. Perhaps Reagan understood that the key to a truly grassroots campaign strategy lay with young people enthused with party doctrine. Perhaps his only audience at that time was an obscure right-wing think tank. At any rate, he got what he was looking for—a public platform to roll out his ideas.

The secret to reaching those voters and thereby winning elections, Reagan argued, was through social issues like "law and order, abortion, busing, (and) quota systems."[16] The strategy was not novel. It was an extension of Richard Nixon's hard hat coalition begun in 1968 and continued in 1972. Nixon and later Reagan noticed that those Democratic voters were slowly switching allegiances based largely on social issues. Wallace's respectable third-party candidacy in 1968 caught their initial attention. Reagan's success pulling off key wins in open primary states in 1976 confirmed that the Wallace strategy was not confined to the man himself.

Early evidence in the wake of Carter's election and inauguration boded well for students of the Wallace school of presidential politics. Republicans picked up quick wins across the country, including a special February 1977 congressional election in Minnesota, which saw a conservative activist defeat an aide for Vice President Walter Mondale in a traditionally Democratic district.[17]

Still, being a Republican in Washington, DC, in early 1977 must have felt like a soldier under siege in the front lines of a war. Democrats controlled all levers of power: the presidency, near supermajorities in both congressional houses, and a working liberal majority on the high court. This was a rare opportunity to truly remake national and international policy in the image of the Democratic Party and working class—so long as they could get along and work well with each other. But, in the words of satirist Will Rogers, "I'm not a member of any organized political party, I am a member of the Democratic Party."

Democrats had struggled to follow through on truly groundbreaking legislation and policies since Roosevelt's New Deal. Democrats may have been effective legislators in cobbling together bipartisan coalitions to pass important legislation, but they were comparatively ineffective in advancing a distinctively partisan agenda. Consequently, they were unable to address key Democratic constituencies like white blue-collar voters with a strong record that spoke to their needs and led to real results. Integration and racial justice would leave a lasting legacy, to be sure. But in the near term, they would not win over paper workers in Wisconsin's Fox River Valley, miners in Minnesota's Iron Range, steelworkers in Pittsburgh, or auto workers in Detroit. A robust jobs plan, updated labor laws, expanded worker rights, plentiful public works projects with job training programs, and universal health care would appeal to those workers. Health care issues were particularly promising. The Democratic Party was able to leverage another social insurance program—Social Security—as a centerpiece of its election campaign strategies and to great effect even though the program itself dated back to the 1930s. There is a reason Republicans have fought Democrats so hard on health insurance reform.

While Reagan and the Republican Party were aggressively putting together a serious plan to take away the white, blue-collar vote from Democrats, the Democratic Party was frittering away its moment.

For Ronald Reagan, these were the years that really mattered, as Craig Shirley explains in his book, *Reagan Rising: The Decisive Years, 1976–1980*. Reagan, the eternal optimist, hoped to rise from the ashes of his 1976 defeat and win the presidency on what would be his third attempt. But to pull it off, Reagan had to endure a grueling schedule and deflect the bevy of attacks from the left and the right. In the lexicon of American politics, Reagan had to be a "happy warrior."

Only a handful of politicians have been granted that sobriquet, including Franklin Roosevelt and Hubert Humphrey. Each waged long drawn-out legislative wars that took years and in some cases a career to win while losing many battles in the interim. Roosevelt had to practically reconstitute the U.S. Supreme Court to keep his legislative achievements intact.

Humphrey, an early champion of civil rights, fought the good fight for sixteen years before Congress passed the Civil Rights Act of 1964. The run-up to the bill signing was long and arduous. Humphrey played second fiddle to Johnson when the Texan served as majority leader. Later, as vice president, Humphrey endured humiliation after humiliation inflicted by the often deliberately cruel Johnson. Humphrey was abandoned in the latter days of his campaign for president in 1968 when he needed Johnson most. He lost to Richard Nixon by less than one point.

To wake up every day prepared to do battle and not be consumed with anger and frustration when one loses or be filled with antipathy toward the antagonists who malign them is a gift. In this way, Ronald Reagan was the counterpart to Humphrey despite being on the opposite end of the ideological spectrum. His patented aw-shucks smile came out of years of practice in front of Hollywood cameras. His folksy one-liners and down-home anecdotes were honed on the rubber chicken circuit ever since he delivered "The Speech" to a national audience and became a high-demand speaker for conservative gatherings. By 1977, Reagan was receiving about three hundred speaking invitations a month.[18]

He used his good humor to persevere through tough times and tough campaigns. Never did he throw an ad hominem attack at his opponents. He didn't need to. His mind, and omnipresent note cards, were chock full of metaphors, riddles, and jokes that always got the best of his adversaries. They were able to accomplish multiple objectives at once, like the line he delivered at the base of the Statue of Liberty in 1980, which brilliantly combined the top issue on voters' minds, their mood at the moment, and a memorable swipe at his opponent: "It's a recession when your neighbor loses his jobs, a depression when you lose your job and it's a recovery when Jimmy Carter loses his."[19]

This was only weeks after Reagan delivered a stem-winder acceptance speech at the Republican National Convention where he laid down a then overlooked passage that would resurface thirty-six years later in another Republican presidential upset: "For those without job opportunities, we'll stimulate new opportunities, particularly in the inner cities where they live. For those who've abandoned hope, we'll restore hope, and we'll welcome them into a great national crusade to make America great again."[20]

It's an article of faith that for a challenger to defeat an incumbent, the latter has to screw up. You can run a flawless campaign but unless your opponent fouls up, hopefully more than once, you simply won't win. A challenger thus has to win two campaigns instead of one. First, they must convince the voters that the incumbent is not doing their job and not worthy of reelection. Then the candidate must give a compelling reason to vote for them. This is one reason of many why it is difficult to defeat incumbents.

At times in 1980 it seemed like Carter was trying to lose reelection. With stagflation rocking the country, Carter refused to reassure Americans. Instead, he blamed them for rising inflation and flatlining growth, a heretofore unrealized phenomenon called stagflation.

Like a tried-and-true neoliberal Democrat who emphasizes free-market solutions like deregulation, tax cuts, and monetary policy over government programs, Carter bought the old canard that higher wages pushed inflation, and that slowing wage growth would solve the economy's fundamental problems.

In 1978, Carter called on Americans "to follow the example of federal workers and accept a lower rate of wage increase," and then tried to shame them into doing so: "We want something to be done about our problems—except when the solutions affect us. We favor sacrifice, as long as others go first."[21]

As a born-again Christian, it was second nature for Carter to craft a speech that sounded like it could be delivered from a pulpit rather than at a campaign rally.

Threading religion into his politics cut both ways for Carter. It gave him credibility with evangelicals, a growing, vote-rich constituency that Democrats had previously ignored, but it also pinned him into biblical bromides that did not always comport with political objectives, such as love thy neighbor, turn the other cheek, and do unto others.

Not all Democrats were neoliberals. And not all Republicans believed in supply-side economics, a theory that argues that tax cuts would raise additional revenue. Enlightened Republican business leaders understood the importance of workers and consumers in a market economy. Since about 60 to 70 percent of GDP is consumer spending, happy workers are productive workers.[22]

Large business-to-consumer Fortune 500 companies joined unions and liberal Democrats in decrying such short-sighted economic policies. This was the odd-political-bedfellows coalition consumer activist Ralph Nader assembled whenever he wanted to drive a stake into the heart of American capitalism. He divided the business world into consumer-facing and business-facing enterprises. The former had brands to protect and customers to keep happy. Like Henry Ford at the turn of the century, these companies understood the value of paying their employees a good wage so they too could afford the products they made. It was a win-win scenario. This philosophy was both good for workers and good for business. These high-wage companies became "safe, stable and perennially profitable, impervious to economic downturns."[23]

The latter group, business-to-business companies, were typically manufacturers that organized under the umbrella of the right-wing National Association of Manufacturers. Unlike long-established companies that did well in good and bad economic times, "they tended to operate with less secure profit margins" and consistently scapegoated government involvement, environmental protections, and unions for any decline in business fortunes.[24] There was one person who could marry the two and still win elections.

Ronald Reagan made working families, particularly down-and-out blue-collar workers, of which there was no shortage in the late 1970s, feel good. At the same time, he slipped past them a regressive and disadvantageous economic agenda. His economic philosophy of tax cuts, hyper deregulation, and controlled money supply helped those who needed help the least. But the top earners were also Reagan's and the Republican Party's essential campaign contributors, an unvirtuous cycle of the first order. In addition, Reagan's rhetorical gifts set him apart from his Republican competitors.

In 1977 and 1978, as Carter was pushing through the Panama Canal Treaty that would return ownership of the land to Panama and as Republicans were falling in line with the president one by one, Reagan stuck to his guns and opposed the treaty to the bitter end. He scored big political points with the Republican base and built up immense political capital that could be cashed in down the road should he run for president. Moreover, he demonstrated to the Republican faithful that there was such a thing as an understanding and sympathetic conservative. You could indeed have it both ways. You could aggressively oppose a seemingly compassionate measure but not come across as angry or hateful.

The contrast between Reagan and the more uncouth Panama Canal Treaty opponents who were "revealing themselves as increasingly nasty and foul,"[25] was refreshing. Maybe Republicans didn't have to settle with warmed-over moderates like Gerald Ford or hard-to-pin-down conservatives like Richard "Tricky Dick" Nixon.

In 1978, Republicans picked up three seats in the Senate and fifteen seats in the House. They also gained six governorships. It was their best year in twelve years, when they had picked up fifty seats in 1966. Of the House seats they won in 1978, five were in the Midwest. One of those seats was Wisconsin's 8th district in the northeast corner of the state, where a young legislator named Toby Roth, a native of Appleton, knocked off Democratic incumbent Father Robert Cornell in the country's second most Catholic district—and by a whopping fifteen points. Cornell was one of the first "Watergate babies" to bite the dust. It was one of the key harbingers of the looming presidential contest in 1980.[26]

Another harbinger that spelled trouble for Jimmy Carter and the Democratic Party occurred in Reagan's California. A statewide referendum called Proposition 13 would cap property tax rate growth at 1 percent plus voter-approved bonds and interest, cap valuation increases to 2 percent per year and require a two-thirds supermajority vote of the legislature to raise state taxes. Reagan was a fierce champion of the measure. It passed by an overwhelming margin, 63 percent to 34 percent. The measure was noteworthy for the severity of its policy implications, the margin by which it passed, and the state in which it was adopted. Californians had a relatively high tolerance for tax burdens. Although it voted consistently for Republican statewide candidates, the state's politics and culture were liberal. But this year was different. Property taxes had been in a sharp upward trajectory and had doubled between 1960 and 1972; meanwhile, anti-tax grassroots organizing was gathering steam,.[27] It was a perfect environment for anti-tax forces to coalesce to defeat liberal interests and Democrats. Reagan could not ask for a better setting to capitalize on his conservative, anti-tax message.

In the 1977–78 cycle, Reagan's political action committee, Citizens for the Republic—a precursor to today's leadership PAC, which nearly every member of Congress has—reported $2,662,746 ($12.5 million in 2023 dollars) raised.[28] It was used to elect Republicans in key districts and states around the country. Its immediate political objective for Reagan was to collect chits for a presidential run in 1980 and to pay for volumes of public opinion survey research to help him construct his own platform and message in 1980.

He and everyone else knew exactly where he stood on the issues. He had made that crystal clear at the 1976 Republican presidential convention, in his two runs for governor, and in "The Speech" back in 1964. Thanks to his nonstop campaigning during the last fourteen years and weekly columns and editorials on the CBS radio network, his name recognition was high. Reagan had a message that fit the times like a glove. He would be a formidable opponent in the coming Republican presidential primaries.

Meanwhile, it seemed Carter's fortunes could not get any worse. To his credit, he collected three major foreign policy wins, passing the Panama Canal Treaty, getting Congress to allow American fighter jets to be sold to Egypt and Saudi Arabia (as well as Israel), and brokering a peace agreement between Egypt and Israel.[29] But Carter could not catch a break on the home front. His legislative agenda was stalling, and the economy was getting worse by the day. Moreover, U.S. Senator Ted Kennedy of Massachusetts, representing the left wing of the party, was in the early stages of testing the waters for a challenge in the presidential primaries against Carter. It seemed that Kennedy was angling to be the Ronald Reagan of the liberal movement of 1980.

Many of Carter's failures on the domestic front were self-inflicted. While 1978 was viewed as a productive Congress for Carter and the Democrats, domestic policy accomplishments were nothing to write home about for working families. One exception was the Humphrey-Hawkins Full Employment Act of 1978, that fulfilled a New Deal–era goal of building in full employment as the guiding principle of U.S. monetary (Federal Reserve) and fiscal (Congress) policies, albeit a watered-down law without teeth. Critics would pan it as symbolism without substance. Not surprisingly the millions of suffering working families felt they were an afterthought because there was little direct benefit to them: corporate tax cuts, elimination of important middle income tax deductions, rejection of most union initiatives, and increased Social Security taxes. To American Federation of Labor-Congress of Industrial Organizations (AFL-CIO) head George Meany, Carter "was the most conservative president since Calvin Coolidge." The United Autoworkers chimed in that they "will tell Ted [Kennedy] to get off his swivel chair and flatly declare himself a candidate against Jimmy Carter for the Democratic party presidential nomination." Carter was only too happy to return the compliment: "[Liberals are] the one group of Democrats with whom I feel uncomfortable. They have a commitment to political suicide in order to prove some far-left philosophical point. It's really disgusting."[30]

Now more than ever, Democrats needed to prove Will Rogers wrong. They needed to get their act together and prepare for the gathering storm.

About twenty minutes north of Gordon Lightfoot's "musty old hall" in Detroit—the Mariners Church of Detroit, which holds a memorial service each year on the anniversary of the sinking of the *Edmund Fitzgerald*—sits Macomb County. If you could drop a pin on a map of the country where Democrats saw their fortunes turn and watched Reagan defeat Jimmy Carter, it was Macomb County. In the postmortem election analysis, pundits and scholars pointed to this northern Detroit suburb as the prototypical county where reliable blue-collar Democrats switched allegiances and voted for Reagan, becoming so-called Reagan Democrats.[31]

In fact, as recently as 1960, Macomb County had been the most Democratic county in the country. This is where most of the auto workers rested their heads and raised their families. It was clear in those days which party was on their side. That year they voted for John F. Kennedy by a margin of nearly 2–1, and in 1964 for the Democratic presidential nominee, Lyndon Johnson, by a margin of 3–1.[32]

From the middle of the 1970s to 1980 the auto industry's fortunes changed. The market share of the Big Three Detroit-based auto companies—General Motors, Chrysler, and Ford—fell twelve points, from 84 percent to 72 percent. At the same time, Macomb County and Michigan overall lost one hundred thousand auto industry jobs. Much of these losses were the result of the inability or unwillingness of domestic producers to keep up with changing consumer demand, as David Halberstam well documents in his epic work *The Reckoning*, which follows the rise of Nissan and the fall of Ford from the post–World War II era to the mid-1980s.[33]

But a lot had to do with American trade policy that continued to favor foreign producers. This should not have come as a surprise to Republicans. In 1977 conservative activist Paul Weyrich and Republican leaders reached out to and met with union leaders in Youngstown, Ohio—an area not unlike Detroit in the 1970s, relying on a strong manufacturing economy—where the Youngstown Sheet and Tube plant closed in September of 1977. In addition to offering the traditional right-wing explanations for economic woes—too much regulation, high budget deficits, government interference—they raised the matter of unfair trade practices, an economic phenomenon that usually progressives emphasize in economic debates.

If Weyrich and company were to be believed, the GOP would be able to make impressive inroads with this hardcore Democratic constituency on Republican issues. On the other hand, Jimmy Carter was a full-throttle supporter of expanded free trade. The cornerstone of his foreign policy was human rights, and Carter used trade as a tool to reward and punish countries according to whether they followed these policies. Combined with his decidedly neoliberal economic policy, this offered Detroit autoworkers little reason to support his reelection in 1980. As the manufacturing base in and around Detroit cratered, so did the fortunes of Democrats.

The follow-on postmortems that occurred in the following decades have been mixed as to the reasons for the defection of these voters. Election watchers in 1980 identified social issues like race, busing, and abortion, which moved voters into the Republican column. Others argued it was the faltering economy that made the difference. American manufacturing had collapsed, eliminating thousands of good-paying steel, shipbuilding, automaking, and maritime jobs. Finally, there was the theory that suggests Reagan's victory over Carter in 1980 was due more to Carter's perceived poor performance in office than Reagan's policy agenda or the emerging salience of social issues. Contrariwise, Reagan's reelection in 1984 was a classic scenario in which voters reward an incumbent for a perceived job well done.

To defer to Stan Greenberg, the celebrated Democratic strategist and pollster who led Bill Clinton to victory in 1992 and who coined the term "Reagan Democrat," economic factors and failed political leadership were the main culprits in 1980.

In 1996, during Clinton's reelection campaign, Greenberg wrote a memo for a public audience, which, with the benefit of sixteen years of hindsight, zeroed in on economic issues as a deciding factor in Reagan Democrats' vote preferences:

> "These were disillusioned, angry voters, but they were not Republicans. They spoke of a broken contract, not a new vision. Their way of life was genuinely in jeopardy, threatened by profound economic changes beyond their control, yet their leaders, who were supposed to look out for them, were preoccupied with other groups and other issues. These voters wondered why they weren't the central drama of the Democratic party. . . . Macomb Country voters knew they were in trouble. With the economic contracting and being reconstructed, so were their dreams. Underlying all the political talk was a lot of simple fear. Givebacks and layoffs, foreign imports and robotics, and the sight of industries moving South—these things now dominated their economic world."[34]

In 2008, in a *New York Times* op-ed column, Greenberg said good-bye to Macomb County as an election harbinger, the phenomenon that made his career. "I'm finished with the Reagan Democrats of Macomb County," he wrote. Instead, he said hello to neighboring Oakland County: "I trust we will see the candidates rush from their conventions to Oakland County."[35]

Greenberg argued that Oakland County better reflected America's changing demography and economy over the previous three decades. Oakland County was more diverse and affluent, and the Macomb County region relied less on the auto industry than in 1980. Said David Dulio, professor of political science at Oakland University, "[Trump] spoke directly to many voters in Macomb County, including trade (i.e., NAFTA) and jobs (especially manufacturing jobs). This clearly resonated with voters in the county."[36]

In Macomb County, just 27 percent of residents hold a four-year college degree.[37] Next door in Oakland County, nearly twice as many residents (50 percent) have at least a bachelor's degree.[38] Oakland and not Macomb was the kind of county Democrats should pay heed to, Greenberg argued.

You could say Greenberg's attitude created a self-fulfilling prophecy. Democrats did not focus on the Macomb counties of America, and not even on the somewhat more affluent Oakland counties of the country. As a result, Democrats' standing in inner-ring suburbs and rural areas eroded to unprecedented low levels. Today, rural America has totally fled the Democrats and is solidly behind Republicans.

Macomb County is no longer considered a bellwether county. In its place are counties in neighboring Wisconsin like Outagamie County (Appleton and Fox Cities) and Brown County (Green Bay) in the northeast region. In two townships in Outagamie County, fewer than one out of every five voters chose Joe Biden in the 2020 election. One of those townships (Liberty) had given Barack Obama 45 percent of their vote in 2008.[39]

But the real change in voting trends in the last decade occurred in small and medium-size factory towns across the country. Between 2012 and 2020, Democrats lost almost 2.5 million votes in these communities, about five times the number of votes they gained in American suburbs.[40] So much for soccer moms. Maybe it's still about the Reagan Democrats. Perhaps Greenberg spoke a bit too soon.[41]

The fortunes of Jimmy Carter and the Democratic Party did not improve after the 1978 midterm elections. The fortunes of Ronald Reagan, however, kept climbing. In the primaries of 1980, he made mincemeat of the other conservatives, particularly U.S. Senator Bob Dole, ex-Nixon aide Alexander Haig, and former Texas governor and Democrat John Connally. Shortly after, the contest came down to Reagan and former CIA director and RNC chairman, George H.W. Bush. Reagan's attractive message, hard work, and discipline on the campaign trail, deep grassroots support, and plentiful resources were too much for the blue-blood scion of the New England Republican establishment.

In addition to his message and finesse on the campaign trail, Reagan used his patented one-liners to devastating effect, always getting the best of his opponents.

At an unruly debate in Nashua, New Hampshire, during the primary, after the moderator refused to allow candidates other than Bush to participate against Reagan—a one-on-one face-off with Reagan would benefit the trailing Bush tremendously—Reagan grabbed the microphone and undressed the hapless moderator: "Mr. Green I am paying for this microphone."[42] The Reagan campaign had paid for the debate because they were concerned that they might run afoul of campaign finance rules.

The crowd erupted in applause and approval. The debate proceeded with all candidates participating. Reagan performed well. But the way and manner he took control of the situation before the debate spoke loudest of all.

Reagan defeated an eight-person Republican primary field with 50 percent of the vote. Bush hung on until May 26 before dropping out of the race; he won six primaries compared to Reagan's thirty-two.[43]

"All in all, it had been a truly bad year," Carter biographer Kai Bird wrote in his treatment of the Carter presidency, *The Outlier: The Unfinished Presidency of*

Jimmy Carter, referring to the year 1980.[44] He was right. One cannot imagine another presidential election in which seemingly everything went wrong for the incumbent. A credible primary threat. A high-profile hostage crisis. An embarrassingly failed mission to rescue the hostages. High inflation. High unemployment. An Olympic boycott. A grain embargo. More oil price increases. Even a volcano eruption.

Without any of the major issues resolved—particularly the hostage crisis—the voters did what dissatisfied voters invariably do: they voted the incumbent out.

Stu Eizenstat, former domestic policy advisor to President Carter (1977–81), accurately and importantly points out in his memoir that neoliberal economic policy and neoconservative defense policy began with Carter—not Reagan.[45] Indeed, it was Carter that accelerated defense spending, deregulated key industries that unleashed competition within the heretofore sacred cow economies, initiated supply-side economic policy, and tapped the brakes on touchstone liberal reforms, particularly universal health insurance (although Carter would do an about-face and call for a national health insurance plan in his one and only debate with Reagan, one week before the 1980 election).

Carter foresaw a political and economic conservative trend within the electorate which he first observed in his retail campaigning in the state of Iowa over one year ahead of its 1976 presidential caucuses.[46] It didn't hurt that he governed a conservative state in the Deep South. Carter adroitly balanced his liberal economic and social philosophy with that of his largely conservative constituency and had a nose to sniff out these trends. In that regard, he was almost alone among the rest of the primary field that consisted mainly of western and midwestern Democrats.[47]

By the end of the 1970s, the kind of political schizophrenia experienced ten years prior—that is, the fall of the New Deal era and rise of the New Right—would be noticeably more nuanced.

The future of American politics would not be a one-dimensional liberal-versus-conservative paradigm. The political pendulum would not simply swing left and right. Rather, in time, it would spin off its fulcrum entirely. While this catastrophe would take several decades to unwind, if one were to closely inspect the gears upon the grandfather clock of American politics, one could notice a screw or two loosening and a gear or two slipping. It began with Reagan and ended with Donald Trump—so far.

The pendulum moved decidedly rightward where Reagan took Carter's conservative policies and pumped them up with steroids. The upper tax bracket was slashed from 90 percent to 35 percent. Defense spending ramped up by 7.5 percent, and by 34 percent and 14 percent in procurement for aircraft and missiles, respectively.[48] And key New Deal and Great Society-era social programs were tossed into the ash heap of budget cuts.

Titular Democratic leader, Speaker Tip O'Neill, put up little fight, acceding to a supposed mandate Reagan received from the 1980 election. Reagan's first budget with its attendant tax cuts would pass the Democratic House easily with the support of sixty-three Democrats.[49] Reaganomics continued full throttle despite an *Atlantic Monthly* exposé in December of 1981 in which Reagan budget director David Stockman told a former confidant, William Greider, "None of us really understands what's going on with these numbers."[50] It was a damning article because Stockman was the brains behind the Reagan budget; its own author was admitting it was nothing more than a fraud.

While Democrats would maintain a robust majority in the U.S. House and flip the U.S. Senate in 1986, the national mood continued moving to the right. Reagan won reelection by the greatest electoral margin in U.S. history. Reagan vice president George H. W. Bush would win a "third" Reagan term in 1988, the first sitting vice president to do so since Martin Van Buren 152 years earlier in 1836. Meanwhile, national Democrats were struggling. Moderate and conservative Democrats established a political organization, the Democratic Leadership Council, which advocated neoliberal economic policies, particularly free trade, deficit reduction, and welfare reform along with a bone or two for liberals such as earned income tax credits for poor people and increased secondary and vocational school funding.[51] Its fourth chairman, Arkansas governor Bill Clinton (1990–91), would use the group as a springboard to realize a boyhood dream.

Economic issues dominated the campaign, which was not a surprise because the country was still struggling to get out of the 1990–91 recession. Powered by a clever campaign haiku that was repeated throughout 1992 and featured the line, "It's the economy stupid," Clinton seemed to have won a mandate to enact an economic agenda that matched the enthusiasm and needs of the decisive voting bloc of Reagan Democrats that had helped secure his victory.[52]

Clinton's words about the economy proved more bluster than action. Whether it was the conservative economic governor of a southern state shining through—organized labor constitutes only 5 percent of the workforce whereas one in five jobs in Arkansas depend on trade[53]—or the neophyte's inability to properly pull and push the levers of power in official Washington was unclear. What was clear was the results. Which left much to be desired.

Tax brackets during the Clinton years remained largely unchanged. Defense budgets were maintained despite the end of the Cold War and the United States claiming the mantle of sole superpower. Working families received token relief. Free trade deals dominated the Democratic president's economic agenda. And the president's first budget emphasized addressing budget deficits and increasing investor confidence in Wall Street rather than putting money in people's pockets. In Clinton's defense, he wasn't entirely sold on his own economic policies. "You mean to tell me that the success of the economic program and my re-election hinges on the Federal Reserve and a bunch of fucking bond traders?" he asked his advisors.[54] The Clinton economic recovery would be top-down and resemble the economic policies of Reagan in 1981 more than Lyndon B. Johnson's Great Society agenda in 1965 or FDR's New Deal. Moreover, Clinton made little effort to revive the shipping industry. His laser focus on trade deals at the expense of building the manufacturing economy was indistinguishable from his GOP predecessor's policies.

The Washington Consensus of neoliberal economic policies, fiscal restraint and deficit reduction remained as such. Neoliberal economic policies dominated national budgets and both Democrats and Republicans agreed to them. Whether it be gridlock, consensus building, or misinterpreted voter mandates, U.S. economic policy remained largely intact for the first decade and a half of the new millennium under the George W. Bush and Obama administrations.

Reclining behind his desk on the twenty-fourth floor of Trump Tower in the fall of 2011, its namesake landlord received a clown car parade of GOP candidates.[55] A nominal Democrat for years, Donald Trump had entered the political arena a couple of years earlier when the former bankrupt casino magnate instigated the so-called birther movement,[56] which cast doubt on President Barack Obama's citizenship. Trump was thinking well beyond 2012 and par for course was thinking well beyond the candidates and the party behind them. He was thinking of himself.

Obama's inability to salve the wounds of the body politic in the first half of his first term left much to be desired for his voters. According to Pew research, the so-called enthusiasm gap between Republicans and Democrats, which measures the difference in how excited partisan voters are about the upcoming election, was 20 percent (62 to 42 percent) in the late summer of 2010.[57] The forty-fourth president's full throttle push for the most comprehensive health insurance program since Medicare unsettled voters. Combined, these two phenomena would sink Democratic candidacies across the country—Democrats would lose over one thousand legislative and congressional seats—and birth a generation of GOP wunderkinds including Wisconsin's Scott Walker, Florida's Marco Rubio, and South Carolina's Nikki Haley, all of whom would eventually run for president.

With a panoramic view of the Manhattan skyline behind him, Trump no doubt wondered, "Why not me?" Unimpressed with the GOP field—he would elect not to endorse a candidate in 2012—he wondered not only if he could run but if this were the right moment. The answer to the first query was yes, the second question, a reluctant no. Something told him timing was not right, particularly for the brand of politics he had in mind. He'd have to wait.

On April 30, 2015, on the lawn behind the U.S. Capitol in what seemed an impromptu press conference that was more an afterthought than a planned rollout, Vermont independent U.S. Senator Bernie Sanders announced his candidacy for the presidency. Former Democratic U.S. senator, secretary of state, and first lady Hillary Clinton had already announced her candidacy and was assured of the Democratic nomination—or so she thought.

Out of the gate, Clinton was leading Sanders by sixty points, 75 percent to 15 percent. By the fall of 2015, Clinton had jumped ahead to a 45–1 superdelegate lead over Sanders.[58]

Meanwhile, gliding down the escalator into the lobby of Trump Towers on June 16, 2015, ironically to Neil Young's rebellious "Rockin' in a Free World," Trump launched his populist campaign for the presidency. Years of honing his ear as a reality TV host and global swashbuckling flimflam man had prepared him for this day.

Despite the opulent chandelier and glow of faux gold-trimmed bannisters in the lobby, Trump made clear he would be the answer to the woes of the working class. He would not be your father's Republican, unlike the other leading candidates, particularly political scion Jeb Bush. Trump would set up shop in the dark recesses of Republican smoke-filled rooms and exploit to the fullest people's base fears against anyone who looked different than his targeted audience, dispirited middle-age white men and women.

From the base of the escalator and in front of his super-model wife, Trump said, "When Mexico sends its people, they're not sending their best. They're not sending you. They're not sending you. They're sending people that have lots of problems, and they're bringing those problems with us. They're bringing drugs. They're bringing crime. They're rapists."[59]

Within weeks, Trump would be leading a crowded GOP and would never surrender his lead for the rest of the campaign.[60] Sanders, on the other hand, would take a little extra time to catch up to Clinton. Despite pulling off twenty-two primary victories, the senator would come up short at the Democratic National Convention on July 26, 2016, and throw his support behind former secretary of state Hilary Clinton.[61]

Lax campaign finance laws and dark money would have little bearing on the Trump and Sanders candidacies. Instead, a disaffected and disgruntled electorate that had had enough of the mainstream elements of both parties and were resolved to do something about it turned the tide toward the populist candidates. The only difference was Trump did better than Sanders and won his party's nomination.[62]

There were differences between the two populist-oriented campaigns, but in many ways they were remarkably alike. Neither candidate was strongly associated with their respective party machinery; in fact, Sanders wasn't even a Democrat and Trump had given thousands of dollars to Democratic candidates and causes over the years.

Leaders within each party detested Trump and Sanders. Arenas packed with crowds swooned over the populist candidates, while their more traditional opponents were confined to meet and greets, modest town halls, and cocktail fundraisers. Neither Trump nor Sanders received much support from big-name donors. Both raised money hand over fist from small online donations. Both took full advantage of the troubling economic and social failures that boxed in party leadership, elected officials, and voters. While neoliberal policies ruled, quality of life and economic opportunity declined. The wealth gap exploded,[63] student debt soared,[64] and infrastructure investments into new roads, bridges, water systems and electrical grids were ignored.

Trump scapegoated the immigrants who flooded the country, committing crimes and taking away jobs, as well as minorities who took advantage of affirmative action policies to receive employment opportunities and college entrance slots for which they were not qualified, he claimed. Sanders affixed blame on corporations who, he argued, bought off politicians in exchange for tax breaks.

The American Populist Party had been founded more than a century earlier, but its strictures were timeless. The Populists rallied around the classic big-guy-versus-the-little-guy and organized-wealth-versus-the-common-man paradigms. Put simply, "Populism is defense of rule by and for the populace—that is, the common people."[65] Varieties of populism could include "New Deal reforms, consumer rage against business, [and] ethnic belligerence."[66]

The historical roots of Sanders's populism were obvious (see above). But what about Trump? He championed tax cuts for the rich and would sign into law Paul Ryan's signature tax legislation in the first year of his administration. Had it not been for John McCain's "no" vote on repealing the Affordable Care Act in the spring of 2017, Trump would no doubt have signed that legislation into law, tossing eighteen million Americans off health insurance.[67] What Trump did do, however, was load up his vocabulary of "tough talk against malevolent elites with ugly scapegoating of marginalized groups" as historian Charles Postel characterized it in the midst of the 2016 presidential contest.[68] In that regard Trump resembled Democratic-populist president Andrew Jackson, as Postel observed. Jackson had referred to the American Indian as "inferior," and lacking "intelligence, industry, (and) moral habits." Jackson advised them to "yield to the force of circumstances and ere disappear."[69] He signed the Indian Removal Act of 1830, which banished Native American tribes into the American interior west of the Mississippi River, paving the way for wealthy landowners to scoop up more territory and enrich themselves.

Sanders would fit in with history's revered prairie populists like William Jennings Bryan. He could easily be identified with the Populist Party in addition to the bimetallism movement and the Greenback Party of the mid- to late nineteenth century. Both of these championed loose monetary policies to get cash into the pockets of working people, particularly farmers, to pay down debts owed to banks and corporations. Trump would be at home with nativist political parties, particularly the Know-Nothings of the mid-nineteenth century. Trump did resemble latter-day populist parties[70] in his vocal opposition to free trade deals and China's encroachment in the domestic economy. It was more than just campaign rhetoric and bluster. Trump would follow through on campaign promises and implement protectionist trade policies that, while boosting certain commodity industries like steel, would crush other businesses further down the supply chain, particularly autos, car parts, and machine tools.

If you could affix a date to the catastrophic collapse of the political pendulum it would be January 6, 2021. Right-wing American populism would unwind like a slow-motion train crash upon which you cannot avert your gaze or an erupting volcano whose fate you cannot escape.

An angry mob of Trump supporters sacked the U.S. Capitol, smashing windows, momentarily taking over the U.S. Senate chambers, cornering congressional staff and their bosses and attacking Capitol police. For the next several years, the FBI would track down the trespassers, bringing them to justice with felony convictions and prison terms. Trump would never concede the election and would spend the entirety of his successor's presidential term undermining and challenging the election results.

Approximately ten thousand rabid Trump supporters participated in the Capitol insurrection. What are the chances that some of them once built Great Lakes ships, forged the steel that created the hatch covers, or cast the iron that made the ships' bells? What drove these self-proclaimed patriots to the brink of civil rebellion and over the edge into the abyss?

Nothing can excuse the actions of the insurrectionists. But it is a fantasy to say that every insurrectionist was a Republican. It stands to reason that a significant number could have been Sanders's voters. In 2016, one in ten Bernie Sanders primary voters chose Trump in the 2016 election.[71] A similar number backed the president in 2020.

In our contemporary politics it has not taken many votes to swing an election. In 2016, eighty thousand votes distributed among three states determined the election outcome.[72] It was 537 votes in Florida that swung the 2000 presidential election to George W. Bush.[73] Twelve thousand votes in Mississippi handed Jimmy Carter the 1976 election.[74] And razor-thin margins in Illinois and Texas made the difference in John F. Kennedy's 1960 victory.[75]

Shaving off just a small percentage of Trump voters could reconfigure our political culture and determine the fate of this country's civil society. After decades of neoliberal economics dominating the public agenda, President Joe Biden has burnished his credentials as the most pro-labor president since FDR.[76] Spending bills like the American Rescue Plan Act have delivered billions of dollars to communities struggling to recover from the pandemic.[77] One-off stimulus checks, funding for childcare and free school lunches have provided working families with New Deal-style relief.

The fate of our industrial economy hangs on a handful of votes. But will voters notice?

CHAPTER 8

The Rise and Fall of the American Shipbuilding Industry

THE BIRTHPLACE OF THE *EDMUND FITZGERALD* COULD HAVE BEEN A MUSKRAT FARM. When Anthony Pessano came to Detroit from western Pennsylvania in search of great riches in the burgeoning shipbuilding industry, he bought a parcel of marshland along the Detroit River on the U.S. side. But before the sale could be consummated, its owner, a self-made muskrat hunter, asked the new owner in his native tongue: "Et si je vends ferez-vous une réservation pour nous permettre de venir sur le terrain et de tirer le râle musqué?"[1]

How Pessano responded to the local Frenchman's request to hunt muskrat on his family's land is lost to history. But if his stomach had any say in the matter, it's certain Pessano acceded to his wishes, because the otherwise troublesome vermin was a regional delicacy not unlike the sumptuous lobster on the East Coast. It also didn't hurt to get on the good side of the locals.

In 1902, Pessano bought the former Samuel F. Hodge and Company near Detroit's River Rouge–Ecorse riverbed area. Pessano immediately went to work converting the old machine and engine works plant into a shipbuilding facility, the Great Lakes Engineering Works (GLEW). His initial investment in the marshland along the river would soon be used to meet the needs of a thriving and expanding business. Pessano was a brilliant entrepreneur. He seemed to always be a few steps ahead of his compatriots. This included local boy made good Henry Ford who would identify the River Rouge area as an ideal location to build his own flagship plant, the eponymous River Rouge Complex, in 1915.

Pessano thought the Detroit area had great industrial potential and considered it "a little Pittsburg growing up."[2] Indeed, the area in and around his plant became ground zero for one of the country's largest manufacturing hubs, headed by the automotive industry, which was on the "leading edge of economic growth of the period."[3]

In addition to running GLEW, Pessano enticed railroad companies to locate to the area. He bought cheap land around his facility and flipped it to the railroads. He made a nice profit and built up the manufacturing base of the region, redounding to the benefit of his shipbuilding concern.

Pessano was the Andrew Carnegie of the shipbuilding industry. In his formative years, he had worked in the mills and foundries of Cressano, Pennsylvania, an hour's drive from the Carnegie Steel Company in Pittsburgh. Pessano dove headfirst into an established market in his adopted hometown of Detroit and quickly took on the older, larger American Shipbuilding Company. His fights were valiant and often successful, even though the competition out-capitalized him by a factor of 21 ($30 million to $1.5 million).

An expert in staving off labor organizing as a leader in the National Association of Foundries, Pessano developed a paternalistic management style not unlike that of Carnegie. He vertically integrated his business right down to the company town near the plant where adequate and attractive housing was available for his employees and their families. Like any good businessman, Pessano had a good eye for details. "I always recommended that our company paint the houses in cheerful colors," Pessano said. "It rests the mind and eases the eyes. You may not think so, but the color of the home in which a man lives has a little influence."[4]

To say Pessano's ships were built to last would be an understatement. One of its first freighters, the *William G. Mather*, was delivered in 1906 and would help carry the U.S. economy until 1996. Its first ship, the RW *England*, was delivered in 1904 and remained in service until 1960. Its second ship, the ferry *Detroit*, plied the lakes until 1990. Yet another freighter, the *James E. Davidson*, sailed for fifty-eight years.

Fourteen years after Pessano purchased the former Samuel F. Hodge and Company, the U.S. government passed the Shipping Act of 1916. It established the United States Shipping Board (USSB). The USSB regulated all aspects of American shipping, in addition to underwriting a significant share of the costs of shipbuilding. In 1917, U.S. shipbuilding employment sat at 50,000 workers. By the end of World War I, it had soared sevenfold to 350,000 workers.

The USSB was World War I's shipbuilding program counterpart to Franklin Roosevelt's arsenal of democracy in the World War II era when the full weight of the American manufacturing economy was converted overnight into a war armaments and equipment-making machine. In this dry run of sorts, between 1917 and 1919, the U.S. shipbuilding industry produced three thousand ships in 150 shipyards at a cost of $5 billion ($110 billion in 2023 dollars).

If today's neoliberals, Democrats, and Republicans alike held the reins of power in 1918, American industry would likely not have been prepared for a second world war. Saddled with an overwhelming surplus of vessels, officials could have disbanded the USSB and ended government support for ship construction and workforce development. Indeed, as the wartime economy wound down and Americans prepared for a time of austerity, many called for the end of the program entirely. But it didn't happen. Our leaders then were wise.

They knew the difference between responding to changes in market demand and the market value of a public good. American workers and industry needed to handle an uncharted postwar economy and fully realize the potential of the technology they created during the war effort to develop peace-time spin-offs in order to supercharge the new economy. Keeping ship workers employed and shipyards busy—despite market demands—would in the long-term benefit all. Should Uncle Sam need to kick in a few bucks to keep it all going, it would be a small price to pay. Indeed, GLEW "might have been forced to close in 1919" had it not been for government support, according to an account of the history of GLEW by the Marine Historical Society of Detroit, Inc.[5] Without that support, would the *Edmund Fitzgerald* ever have existed?

The workers of GLEW did not put their future entirely in the government or anyone else's hands; on their own initiative, they purchased one-third of the company in 1920. Their move was a precursor to today's Employee Stock Option Plans (ESOPs) but with one notable difference. Today's ESOPs are typically constituted as an option of last resort when no one else wants to buy a company and its jobs are about to head overseas. The GLEW workers' 1920 purchase could have been chalked up to desperation, or an enduring sense of patriotism, or even a savvy business move—or all of these. In any event, it kept the company strong, allowed thousands of hardworking folks to remain on the payrolls, and helped stave off a postwar recession.

On the eve of the Great Depression, GLEW received its 129th ship order. Not bad for an old rusted-out engine works company on muskrat hunting grounds with a fraction of the capital that the market's Goliath possessed. It still played second fiddle to the American Shipbuilding Company, but GLEW's contributions to the industry were many, including building the Great Lakes' largest carrier as of 1959, the *Arthur B. Homer*. It eclipsed by just one foot the previous record-holder: the GLEW-built *Edmund Fitzgerald*.

Its overall success no doubt exceeded its overachieving industrialist Pessano's own expectations. Pessano passed away in 1923. Fortunately, he had built up a strong cadre of front office leadership and a workforce second to none. Being in one of the country's top manufacturing workforce pools—a result in part to Pessano's leadership in developing the local economy—also ensured the company always had access to a wide and deep pool of skilled labor. Pessano's early partner, John R. Russel, took the helm in 1917 and steered GLEW through the early years of the Great Depression.

The company won just one order between the laying of the *Eugene P. Thomas* of 1930 and the end of 1938. By way of comparison, GLEW had filled fourteen orders in the previous nine years (1920–1929) and a staggering twenty-six orders delivered in 1919 alone.

And like a good neighbor, Ford Motor Company threw some business GLEW's way during the Great Depression to keep them afloat. GLEW received contracts for four canal cargo vessels, one tugboat, a yacht, and "several contracts" to build out Ford's barge fleet.[6] In addition to the modest order list, the shipyard kept itself busy with repair and maintenance jobs.

As it is said, it takes one to know one. Ford likely found kindred spirit in Pessano and his investors. Ford appreciated the symbiosis of a strong economic ecosystem. This is something Pessano also prioritized when he spent time in his early years attracting outside investment from the railroads and making space for Ford to grow his company in the River Rouge-Ecorse area.

In 1936, the U.S. Shipping Board disbanded and Congress passed the Merchant Marine Act in its place. The act established commercial shipping's new regulatory body, the United States Maritime Commission. The legislation's priority was to rebuild the country's dilapidated shipping fleet of the World War I era with a ten-year five-hundred-ship cargo vessel order. In addition, the act created a program that would pay the cost differential between a foreign shipyard and a domestic shipyard for commercial vessels.

The new law would update the country's commercial ship fleet and create much-needed capacity to haul iron ore from Minnesota and the Upper Peninsula of Michigan to the hungry steel mills of Ohio, Indiana, Pennsylvania, and throughout the Midwest. This in turn fed the shipyards, machine tool factories, armaments factories, and vehicle part foundries that were building America's arsenal of democracy. By 1942, Great Lakes ships would be moving eighty-eight million gross tons of ore. One decade earlier, ships were moving just *4 percent* of that total, or 3.5 million gross tons.

GLEW filled three Laker orders and six of the sixteen orders for the Class Lake Bulk Freighter program established by the government. The balance of the Laker order went to GLEW's longtime competitor, the America Shipbuilding Company.

The Class Lake bulk freighters were monsters. They were each 620 feet long, in the same ballpark as the *Edmund Fitzgerald* (729 feet). Combined, the sixteen cargo vessels would meet the country's steel mill demand. By 1943, every five days—the approximate trip length for freighters moving from Minnesota to Ohio—the sixteen ships would move almost a quarter of a million tons of ore. Every military unit, from the English Channel to the seaways of the South Pacific, relied on Great Lakes cargo vessels to help make the aircraft, vehicles, equipment, and material they needed to fight and win the war.

At the conclusion of the war, American shipbuilding had done its job and done it well. And just as the world's diplomats and politicians would focus now on winning the peace, so too would American industry focus on the economic equivalent, ensuring a strong manufacturing base across all industries: steel, toolmaking, automotive, paper, and of course shipbuilding.

Many soldiers on the front lines returned home without a job or even a place to rest their heads. Employment prospects were nowhere near where they were when war raged in Europe. America was struggling with a housing shortage and entering a period of austerity, just as it had at the conclusion of World War I. With the war machine screeching to a halt and the job market flooded with sixteen million returning veterans, it would be a time of great social, cultural, and economic tumult.

Political leadership was lacking, too. Truman oscillated between support of labor and business during his administration. Sometimes he was with the worker. But sometimes, in some of the more high-profile political fights of the postwar years, he was not. Said Walter Lippmann, the leading political commentator of the time, "I never thought that his quick way of shooting from the hip was the way the Presidency should be conducted."[7] The steel plant seizures, which of course business did not support, are an example of support for workers, since Truman wanted the owners to accept a relatively labor-favorable wage deal, as per the ruling of a Wage Stabilization Board that he had appointed.

In 1947, Truman vetoed the Taft-Hartley Act, anti-union legislation that had been spurred by a wave of strikes across the country and egged on by conservatives in Congress. Workers were asking for a bigger slice of the pie after sacrificing so much on the battlefield and working so hard on the assembly lines and in the factories during the war effort. Private industry would have none of it, despite profiting mightily from their labor in the war years. Congress overrode Truman's veto; the vote included 126 Democrats of both houses.[8]

When steelworkers struck after negotiations with management for wage increases broke down in 1952, Truman sent them back to work by nationalizing the steel industry; the U.S. Supreme Court struck down the order two months later in the *Youngstown Sheet & Tube Company v. Sawyer* decision. Purchasing power had fallen by one half between the end of the war and the strike. Workers needed a break. Whose side was Truman on?

Not only did Truman fail to put into place a coherent economic policy, but private industry was also equally negligent. Instead of working together for a strong manufacturing base, leaders seemed more concerned about their own businesses. Overall industry trends spoke for themselves. Manufacturing as a percentage of gross domestic product (GDP) peaked in 1947 at 33 percent. By the end of the 1960s, it sat at 27 percent. When the Fitzgerald sank in 1975, it was a mere 22 percent, a one-third decline since the end of World War II. Cornerstone industries like automotive showed signs of waning due to foreign competition, poor trade and economic policies, and anemic responses to changing consumer and market demands.

For a country that had pulled together a war effort in a way totally unmatched at any other moment in history, it was breathtaking to watch its sense of community and patriotism fall apart so quickly.

Nonetheless, manufacturing still offered wages above the norm and employment grew at a steady pace. In other words, despite early signs of trouble, manufacturing remained strong and promising. All that was needed was sound political and industry leadership. It would be some time before America could pen industry's obituary.

"It is important to bring home to our people that the rest of the world pays little attention to maintaining a free market."[9] For years, organized labor and liberal politicians have been screaming this warning in hopes someone from the distant past might hear it and pay heed. It turns out those very words originated from the past and it came from an unlikely set of lips: Basil Harris, chairman of the board of U.S. Lines, the country's largest passenger-freight liner company in the immediate postwar era. Harris, who was also a leader in the establishment of the merchant marine, well understood the importance of strong industry. He believed a sound manufacturing economy was as important and patriotic as readied armed forces. Chief among industries was shipping, a basic and essential industry that facilitated commerce.

Just as industry leaders of today including the ultra-conservative National Association of Manufacturers (NAM) dismissed concerns over expanding trade deficits, shrinking manufacturing base and bad trade deals, so too did less enlightened industry leaders of Harris's time, which also included NAM.

The only thing keeping American shipping in business was government subsidies. For or against the policy, it was an economic fact that both the left and right agreed upon; the only difference of opinion was whether saving and keeping the shipping industry was important or not. It would not be until 1981 that Washington would give its final answer, but in the meantime sensible policy makers, progressive industry leaders, and most if not all labor leaders ensured America's commercial fleet would remain intact and competitive.

Reflecting on the Merchant Marine Act of 1936 in 1945, Basil Harris observed that 100 percent of the subsidy went to workers because the difference in price between the United States and foreign fleets were labor costs. "[Foreign nations] do not build or sail their ships under wage and working conditions which we have established as a fair standard of living. Wages on foreign ships are low, quarters are cramped, food is poor," Harris said.[10]

The only argument against Harris's position and others of like mind was the classical economic argument of comparative advantage: countries should focus on what they do best and to the exclusion of other goods and services. In this example, the thinking went, it was more efficient for countries with lower standards of living to ship cheap goods than for advanced countries like the United States to do so. Consumers would get less expensive products, and the United States could focus its abundant resources and time on bigger and better things. This train of thought would seep into free trade negotiations like the North America Free Trade Agreement of 1993, the World Trade Organization agreement of 1995, and the Central America Free Trade Agreement of 2004. Together, these expanded free trade pacts decimated whole industries and killed tens of thousands of good-paying jobs, particularly manufacturing jobs in Midwest states like Wisconsin, Michigan, and Ohio.

Like elected leaders in the wake of World War I, Harris believed it made more sense for the government to strengthen and maintain a key industry, in this case the linchpin of national and international commerce, than to save a few taxpayer dollars, allow the industry to wither on the vine, and empower other countries to control seaway commerce.

In 1950, the U.S. government disbanded the U.S. Maritime Commission, the body that oversaw implementation of the Merchant Marine Act of 1936 and replaced it with a less robust subsidy program administered by the U.S. Maritime Administration (MARAD).

The year the Fitzgerald sank, the Government Accountability Office, in a report to Congress, encouraged the use of public subsidies to purchase foreign-built ships for U.S. flag operators rather than construct U.S.-built ships. There was also a matter of too many jobs, this in a time of recession.[11] The reason: foreign governments could build ships cheaper and just as well. Their recommendation was akin to encouraging a U.S. consumer to buy a Toyota or Volkswagen instead of a Chevy or Ford because the foreign vehicles are cheaper and there are not enough domestic workers to build American cars.

> Increased merchant ship construction activity in U.S. yards was of growing concern to the Navy. According to a January 1974 Navy Report, as the commercial workload increases, competition for available shipyard facilities and skills tends to increase both costs and delivery times for military ships.[12]

But this was no reason to abandon domestic industry for foreign producers. It would create a vicious downward spiral and unvirtuous cycle where American industry further eroded, and more good-paying jobs were sacrificed.

In 1981, the beginning of the Reagan administration, the United States ended its practice of giving shipbuilding companies construction differential subsidies

(CDS), which were payments of up to 50 percent of the construction cost difference between the U.S. producer and its leading foreign competitor.[13] Worse, the administration had acted unilaterally and hadn't bothered to ask other countries to end their subsidies, too. Thereafter, it was impossible for U.S. companies to compete globally.[14]

According to a study conducted by Marine Insight in 2020, there was not a single American shipyard among the top producers in the world.[15] South Korea claimed the top four positions, and Japan boasted three shipyards. Even Russia ranked among the top ten; its USC United Shipbuilding Corporation placed seventh overall.

The top global producer, Hyundai Heavy Industries of South Korea, generated $34 billion in revenue in 2016. By way of comparison, the total U.S. shipbuilding market size was $35 billion in 2023.[16]

As U.S. shipping slipped into oblivion, foreign producers stayed afloat and surged as their governments pumped billions into them. Not surprisingly, this divergence occurred shortly after the United States ended its construction differential subsidy program in 1981.

Between 2010 and 2018, the Chinese government injected $132 billion into shipbuilding subsidies for Chinese firms.[17] A new program developed in 2022 in India would support the construction of at least fifty new vessels.[18] Part of the government effort would include a $123 million maritime development fund. But India was not doing this for the sake of one-upping a fellow BRICS country (Brazil, Russia, India, China, and South Africa) or even creating jobs but saving money. The country could save $25 billion in foreign exchange if they built up their own commercial fleet. It was a sad commentary when seemingly every other country, including developing nations, understood the value of a strong shipbuilding sector and a reliable fleet for the economy. The United States has gone without a serious shipbuilding program for forty-three years.

Virginia is a right-to-work state and has been since 1947. The Taft-Hartley Act of that year allowed states like Virginia to require unions to recertify annually and required the union itself to collect dues without the benefit of automatic paycheck deduction in the manner of health insurance or a 401k contribution. Virginia has one of the smallest organized workforces in the country.

To maintain a union, members must work extra hard to keep it intact. This means holding annual elections, going door to door, or manning a parking lot to sign up members instead of holding one election and then collecting regular dues deductions from paychecks. It is an onerous burden that establishes a double standard. When an employee signs up for a retirement plan and agrees to deduct a regular contribution from their paycheck, they do not have to re-up every year, let alone win an election to keep their benefits.

Since Michigan became a right-to-work state in 2012, private-sector union ranks have fallen by 143,000.[19] In 2011, Wisconsin adopted Act 10, which effectively ended public-sector unions. In 2015, it implemented right-to-work legislation. Between 2010 and 2018, the Dairy State lost 137,000 private-sector union workers.[20]

So, when the ship builders of Newport News Shipbuilding voted to organize with the United Steelworkers (USW) in January of 1978 and won, it was big news but not necessarily a surprise. The workers of the new union, Local 8888 of the USW took great pride in their work. How could they not? They were the men and women who built the backbone of the U.S. Armed Forces. Their imprint as welders, painters, pipe fitters, and ironworkers would live on for generations. Their skilled craftsmanship was on full display when their masterpieces sat on world stages, as the USS *Missouri* did at the end of World War II or when the USS *Hornet* received Neil Armstrong, Edwin E. "Buzz" Aldrin Jr. and Michael Collins after their splash down in the Pacific following the first lunar landing in 1969.

Winning an election under the auspices of the National Labor Relations Board (NLRB) was not enough. Huntington Ingalls Industries (parent company of Newport News Shipbuilding) refused to recognize them, and the workers struck one year later in 1979. Lane Windham's *Knocking on Labor's Door: Union Organizing in the 1970s and the Roots of a New Economic Divide* gives an exhaustive account of this watershed membership drive that was emblematic of the hurdles labor faced, not just in shipping or in the south but in all industries and in all parts of the country. Yet despite these obstacles, the 1970s proved to be "the first days of a reshaped and newly energized American working class" because the number of workers voting in NLRB elections remained at a similar level in prior decades traditionally viewed as the "heyday of American labor."[21]

On what would be known as Bloody Monday, April 16, 1979, Virginia Governor John Dalton ordered national guardsmen to Newport News to bust heads and show the strikers who was boss. Local police soon joined in, and it became a "war zone," according to Michael Fabey, author of *Heavy Metal: The Hard Days and Nights of the Shipyard Workers Who Build America's Supercarriers*. One of the lead organizers, Bill Bowser, would suffer a permanent limp from the fight.

Through it all Bowser never lost his composure and never lost sight of the necessity of solidarity and the promise of a union. Fabey said:

> A white van pulled up, and a handful of police officers jumped out. One swung a three-foot baton into Bowser's left leg, and he crumpled to the ground. Police descended on him, kicking and clubbing. He folded himself into the fortress of his own thoughts. *Be like Martin Luther King Jr. Nonviolence. Nonviolence.* The beating continued. His body went number. *Please stop. Martin Luther King Jr. Nonviolence. Please make it stop. Make it . . .*[22]

Not until one year later in March of 1980 would a federal appeals court recognize the union's right to organize. Shortly after, Local 8888 voted 4,939 to 1,646 to approve their first contract. It was the conclusion of an eleven-year campaign that had begun in 1969 when Bowser had met with national union leaders to discuss the possibility of organizing the shipyard.[23] Martin Luther King would have been proud.

Today, Local 8888 is the third largest union local of the United Steelworkers, the country's second largest industrial union. The Newport News Shipyard is the largest private shipyard in the country. It is also Virginia's largest employer.

In January of 2019, the company received a two-aircraft-carrier contract for $14.9 billion, the first time that the United States government issued a multi-year contract for carrier construction in four decades.[24] It was all thanks to the vision, hard work, and unbeatable tenacity of Bill Bowser and his union brothers and sisters. The Newport News Shipyard story was not just a classic American story of ordinary people coming together and taking on great odds to accomplish big things. It was proof positive that our nation's best shipbuilding days were not behind us.

Conservative groups like the Heritage Foundation or periodicals like *National Review* are usually not keen on government subsidies. Rather, their members and writers lean on market-based solutions with limited if any government support. In the words of conservative hero Ronald Reagan on the occasion of his first presidential inauguration, "Government isn't the solution to our problem; government is the problem."[25]

You'll have to hand it to these conservative groups. They were consistent. Despite recognizing the critical importance of the shipbuilding industry for national security and defense preparedness, neither Heritage nor *National Review* championed any kind of government support. They wouldn't blink at an increase in government spending for defense contracts with private builders, but when it came to workforce development or leveling the playing field with foreign competition, they offered no solutions, only pablum. "American shipbuilding must be revived. To do this, America must once again be globally competitive, achieved by leveraging novel technologies that address obvious supply chain weaknesses. Such innovation speaks to American strengths."[26] At *National Review*, they believed a landmass alone could get the job done:

> At this point in its history, the nation should take stock of its unique geostrategic location in the northern, resource-rich portion of the Western Hemisphere. . . . Recognizing this, the country should commit itself, as its founders originally did, to taking full advantage of its geographic situation and implementing a comprehensive diplomatic, economic, and military strategy that focuses primarily on securing and stabilizing the four great global commons—the seas, air, space, and cyberspace—as a means of preserving the nation's preeminent position.[27]

If only it were that simple.

CHAPTER 9

Christmas in July

THE SAINT MARY'S RIVER IS THE LINCHPIN OF THE UPPER AND LOWER GREAT LAKES. Behind it sits the Soo Locks, gateway to Superior. Ahead the river flows into the second largest Great Lake, Huron. If you're up for a half-day cruise you can make it to the tip of Lake Michigan at Mackinaw Bridge to the south and southwest.

Just outside the Sault's city limits—as she is affectionately known locally—and tucked behind a small cluster of like homes at the end of a driveway the length of a football field, sits Tom and Chris Farnquist's home, a two-story ranch with an expansive front yard view overlooking the up- and down-bound shipping channels of the lower Saint Mary's River. If anyone deserves the peace and quiet of such a bucolic setting in retirement, it's Tom Farnquist.

No one individual alive has visited the Fitzgerald wreck more than Tom Farnquist and no single individual has contributed more to unlocking the wreck's mysteries. Like the river behind his living room, Farnquist served as a linchpin for the divers, politicians, family members, and history buffs seeking answers to the ship's demise.

Farnquist's exploration began with a just-missed opportunity in 1980 when world renowned oceanographer Jacques Cousteau's son Jean-Michel and his team dove the wreck, the first ones to do so since the iron ore tanker sank five years earlier in 1975. It wasn't much of a dive. The ship, *Calypso*, was literally in the neighborhood and had a few hours to kill before they were expected "in the lower lakes someplace," Farnquist told me.

Despite the notoriety of the legend of the Fitzgerald—thanks in part to the monster success of Lightfoot's hit ballad and the continued finger-pointing and blame-placing among families, the government, ship companies, the U.S. Coast Guard, and others—Cousteau had an eye for ecology, not shipwrecks, and certainly not political controversies. "The Fitzgerald was no big deal to them. They're more into the ecology than anything else," Farnquist told me.

It was unfortunate that Cousteau couldn't spend more time on the dive, because the "battery crapped out" in his diving vehicle the *Sou Coupe* ("diving saucer"), as Farnquist indelicately put it. Farnquist, next in line to dive, would be unable to realize his dream.

Maybe it wasn't meant to be? In a curious twist of fate, the *Calypso* had been damaged by a storm on the Great Lakes. "They [Cousteau and his crew] hadn't seen anything like it before," Farnquist told me. "The waves on Superior are close together, they're used to these rolling seas in the ocean. Well, the waves on the Great Lakes build up, you know, bang-bang-bang beat the hell out of you." It was a fact Farnquist knew and understood too well.

The visit was also unfortunate because Cousteau was pressed with questions by the world media, including the question of why the Fitzgerald sank. The dive was well publicized, and media as far away as Ireland and Australia were blowing up the Calypso's switchboard to be the first to report on the dive. Cousteau felt compelled to answer the questions despite having zero expertise on the matter. Evidently a media relations advisor was not on hand. "Cousteau said it broke on the surface," Farnquist told me. "He said it broke in half on the surface and fell to the bottom. Which I would totally disagree with."

In Cousteau's defense, it was an understandable conclusion. The earlier and well-known November 1958 sinking of the *Carl Bradley* on Lake Michigan broke the ship apart on the surface,[1] although according to former Great Lakes sailor (1976 to 1987) and Green Bay attorney Tim Nixon, nearly every major shipwreck either fell to the bottom and broke apart there or stayed intact. There were two survivors from the *Bradley*, so it could be deduced that it broke apart on the surface. It would follow to reason that a similar ship caught in similar weather conditions would meet its fate in a like manner. Because he had been the only eyewitness heretofore of the wreckage and his dive would be the only visit for nine more years, Cousteau's theory would hold currency until 1989.

In addition to his central role in the legacy of the Fitzgerald, no one individual outside the families of the Fitzgerald has done more than Farnquist to bring closure to the Lakes' most notorious sinking while pulling back the curtain that hides the mysteries of the *Edmund Fitzgerald*'s demise. It was a role he would fill twenty years after the wreck, in July of 1995. Until then, for Farnquist, diving the Fitzgerald and trying to unlock the mystery behind its sinking was all business. That would change after Farnquist made his first of four dives to the wreckage.

The late Fred Shannon of Mount Morris, Michigan, was a retired law enforcement officer turned shipwreck buff. But the term "buff" would do his cause an injustice. Shannon spent his savings to fund an expedition to the Fitzgerald in July of 1994.[2] It would be just the second crewed visit to the wreck since Cousteau's visit fourteen years earlier. His reason for making the dive was not unlike that of the legions of Fitzgerald aficionados. "I've always wondered, it has always bothered me, why did this ship sink?" Shannon said in one of his last interviews before his untimely death in November of 2022.[3]

Fred Shannon was a colorful character, to say the least. Farnquist described him as a "showboat" who went to a press conference and town hall on his dive clad in "a shirt unbuttoned down to a gold necklace with cowboy boots and makeup." Or perhaps more simply and diplomatically, a "renegade," as Great Lakes Shipwreck Society development director and Fitzgerald diver Sean Ley put it.[4] Scott Purvis, who was in Shannon's party and would, at age 12, be the youngest to dive the Fitzgerald, judged Shannon like any preteen boy would. "He was cool. He had a submarine," Purvis told me.[5]

In July of 1994, Shannon rented a tug from Scott's father, Jack Purvis, the largest tugboat operator in the region. It didn't come cheap, and it didn't come without strings attached. In exchange for the rental, Shannon had to take along Scott. This would prove significant because Scott was the first team member who spotted the body outside the wreckage on the floor of the lake. Scott was the first eyewitness to come forward with information about the second body. The significance of this discovery cannot be overstated, and this book marks the first record of it.

A follow-on documentary released by Shannon in 1995, *Expedition '94: To the Fitzgerald!* in time for the twentieth anniversary of the sinking, was intended to turn heads. He wanted to get something with his life savings, and that figure did not include additional funds kicked in by various investors, some of whom did not get their money back when a book failed to come of the dive.[6]

The production featured footage from the dive as well as interviews of former *Arthur M. Anderson* captain Bernie Cooper and former Fitzgerald sailor and whistleblower Red Burgner. But the real selling point was supposed to be the release of the footage taken by Purvis of the body near the wreck. The families would have none of it. They fought back and got a law passed in the Michigan legislature to ban the release of such photography.[7] The photo was never released. It was the first public fight in which the families proved their mettle and demonstrated their resolve and power to protect the Fitzgerald and their loved ones memories at any price.[8]

It was too bad that the documentary didn't get much traction beyond the controversy over its release. Both Cooper and Burgner's interviews shed new light and offered new insights on firsthand accounts from the night of the sinking and the history of the ship.

Overlooking important and compelling new information is a recurring phenomenon in the canon of Fitzgerald research and writing. If you look close enough, you can catch glimpses of gems of fresh insights and revealing information among each documentary, slide show presentation, book, TV show, news article, and magazine feature that has accumulated over the years. Yet little effort has been made to connect the dots to render a holistic narrative of the wreck and the surrounding folklore of the Fitzgerald.

This is not to say Fitzgerald aficionados are incompetent and cannot connect the dots. Quite the opposite. Followers of the Fitzgerald folklore exhibit an almost messianic pull to the folklore, one that impels them to attack the subject like detectives or cult followers. Sault Ste Marie native and retired engineer of the Nuclear Navy, Mike TenEyck, explained the motivation behind his interest: "How do we get the truth?"[9] TenEyck wasn't the only one. He sounded a lot like Shannon: "It has always bothered me, why did this ship sink?"

Fitzgerald folklore is rooted in a wide and disparate diaspora. It consists of brokenhearted wives and melancholic sons and daughters, as well as run-of-the-mill history buffs, maritime enthusiasts, and tourists. Each is attracted to the folklore of the Fitzgerald for their own special reason, including myself. Some appreciate the diversity of interest of their fellow adherents, others do not. This explains the tension between the family members for whom it is all too personal and the layperson who can come across as a Trekkie or, in today's lexicon, a Swiftie for shipwrecks.

The common thread of the Fitzgerald does not resemble a normal diaspora, where an uprooted people who share a common history, religion, or culture spread to all corners of the earth. The common thread of the Fitzgerald is not as strong. This is due in part to the visceral emotion it engenders, fueled by Lightfoot's song that has memorialized the wreck in the public consciousness. It helps explain why out-of-the-woodwork types like Shannon upend lives and empty bank accounts to satiate their curiosities.

"It's so quiet and dark down there it reminds me of a church knowing that God's there," said then twelve-year-old Scott Purvis of Sault Ste Marie, Ontario, in the aftermath of the Fred Shannon dive to the Fitzgerald in July of 1994.[10] He would soon find himself an accomplice to a controversy: the photograph showing human remains near the wreck of the Fitzgerald. Scott was the one who found the body.

"I had the opportunity to go down for a short time at the end of a day of diving," Purvis recalled to me twenty-nine years later.[11] "The batteries of the sub were getting low not dangerously low or anything, but it was affecting the equipment on the sub. When we went down, we landed on the bottom. We were having trouble knowing where we were on the bottom. We were just kind of doing circles and figure 8's and trying to figure out where we were on the bottom," Purvis told me. "We were [then] noticing clay, heaps of clay pushed up from where she [the ship] hit the bottom.... It kind of told us that we were getting in the area of the wreck."

Farnquist had seen the same thing from his sonar when he made the dive a few weeks earlier on July 4, 1994. "It literally plowed a 100-foot trench. It plowed the mud like a snowplow would plow snow down a road just up on both sides like that," Farnquist explained to me.[12] His two hands came together and dovetailed outward like he was pushing the windrow of snow himself. "It dug down so deep below the mud that it had hit bedrock—BOOM. And that's what stopped the ship."

It had been twenty-nine years since Purvis made his dive. He could recall his trip with impressive detail, too. "We were seeing things like coffee cups and eggshells, things like that on the bottom," Purvis described to me. His voice was plain and monotonal, just as one would expect from an even-keeled Canadian. But to me his words were vivid and enthralling. In my mind I could see shards of shattered Buffalo China, the kind you would find at a twenty-four-hour diner, sticking out of the clay ground. And playing cards, the ones that Wheelsman Red O'Brien would deal out in games of gin rummy with his shipmates, scattered on top of the silt sand.

Unlike the lower Great Lakes, Superior was free of the high-calcium levels and invasive species that ate away at underwater wrecks. Wrecks and other remnants of human activity found on the bottom of Superior were typically in good condition regardless of age. "The paint [on the ship] was pristine. Still, 19 years later underwater, like it was just painted yesterday," Scott said to me. His recollection matched that of Farnquist almost word for word. Farnquist recalled to me a conversation he had with Cousteau's team following their trip to the wreck in 1980. "They were so mesmerized by this massive ship sitting on the bottom with the painting look[ing] like it was *painted yesterday*."

While the thought of exhuming a body repulsed me, like a train wreck or car accident—no matter how horrendous or graphic—one's curiosity seems to always get the best of them. There's a reason the ten o'clock news leads off with murders, homicide court cases and gruesome car accidents. You can't help but look, or in the case of a Fitzgerald memory, listen and imagine. I was no exception. "And we were coming along, and we saw something up ahead that kind of resembled somebody's bottom end," Purvis recalled to me. "So, as we got closer, we realized there was bone and stuff like that. It was evident that it was a body, so we got up closer, trying not to disturb it."

Purvis continued. "We examined it for quite a while, and we [soon] realized that it [the body] had a life jacket on." Purvis explained to me how it was possible that a life jacket would not only not float but could drag a person further down into the water. "After a life jacket is submerged under 200 feet or so the pressure squeezes the air out of the cork bricks. So, it would actually act like hardwood blocks on your back after 200 feet." Life jackets of that time floated by virtue of corking inside the outside material of the life jacket. "There was a stainless-steel D-ring in the center of the back.... It still had clothing and hair and things like that, kind of a fleshy skeleton."

There was more. Much more. Up until this point, it has only been reported that one body was found outside the wreck of the ship. When asked by an audience member at a book talk in February of 2023, former TV reporter and documentarian Ric Mixter answered the question "Were there other bodies found on the bottom?" by asserting there had only been "rumors" of a second body.[13]

My phone call with Scott would upend thirty years of Fitzgerald lore. "And then we carried on a little farther after that," Purvis continued. "And actually, I think, we saw another one, kind of similar in layout of a body on the bottom. We didn't get any video or pictures of it because it was just kind of on the way by."

I hadn't been a big Fitzgerald disaster buff, and this project was focused on the economic and historical context of the ship. But still I knew this moment carried a lot of significance and meaning to the canon of Fitzgerald research. Even though Purvis didn't or couldn't go into more detail on this particular memory, he said enough for me to appreciate its significance.

While Shannon was organizing his expedition to the Fitzgerald, Tom Farnquist as executive director of the Shipwreck Museum was putting together his own trip to the wreck. Whether one was in response to the other is not clear and whether there was a legitimate rivalry between the two is not apparent either. "Fred Shannon was kind of an archrival of mine but only because the press made it that way," Farnquist told me.[14] Suffice it to say, the twin explorations provided plenty of drama for ship adventurers and Fitzgerald family members, and fodder for local media.

Dr. Joe MacInnis, Canadian physician-scientist and recipient of the Order of Canada for his work in undersea science research in 1976, reached out to Farnquist to make an expedition to the Fitzgerald. Farnquist was viewed as an authority on diving the Fitz—his reputation and work preceded him, and he was a natural to lead a trip.

"He contacted me, knowing that we had got this great 3-D imagery of the wreck from the ROV dive in 1989," Farnquist told me. Farnquist was referring to his first visit to the wreck. The trip was a huge success. Farnquist's party nabbed incredible photographs with remarkable detail. Camera technology was considerably better than nine years earlier when Cousteau did his dive. Moreover, Farnquist's trip was extensive; he captured footage of other parts of the ship and not just the pilot house as Cousteau did in 1980.

The photos had been in high demand. "Bob Ballard wanted to buy the 3-D imagery from us, but we said no," Farnquist told me, referring to the man who discovered the *Titanic* on September 1, 1985. The Shipwreck Society had other plans for developing a program for the museum with the material.

The photos were not for sale. That didn't mean Farnquist wouldn't be up for another trip to the Fitz. MacInnis was not one to accept no for an answer. He pressed Farnquist. "Do you think it'd be safe to put a manned submarine down on it even though Cousteau had done it?" MacInnis had asked Farnquist. Farnquist was fairly confident he could do it. "Yeah, I think so," he recalled to me. And above all Farnquist could do it safely.

Getting Farnquist to say yes was the easy part. The hard part was persuading Harbor Branch Oceanographic to make the trip. Harbor Branch Oceanographic was the marine science institute that MacInnis was working with for the project. They were just one of three institutes in the nation with a submersible vehicle, like the *Calypso*, to make the dive.[15] Harbor Beach Oceanographic was well respected and in league with Jacques Cousteau and his eponymous society. At first it seemed unlikely. Harbor Branch had suffered the loss of two divers in their dive of the *Fred T. Berry* in 1973, including the son of Harbor Branch Oceanographic's co-founder, Edwin Albert Link.[16] "On that trip, they drew straws, and the son of the owner got the after cabin," Farnquist told me. "The sub got tangled up. The son ran out of oxygen."

To say they were a bit gun shy was an understatement. That's why Harbor Branch Oceanographic had to be convinced beyond a doubt that the dive would be safe. MacInnis knew Farnquist's reputation in the diving community was solid and figured he'd be his best chance to get Harbor Branch Oceanographic on board. He was right. "I said, 'I think it is a very very safe wreck to dive because there's not cables all over the place that you could get tangled up in, unless you just recklessly drive around like bumper cars," Farnquist told me. The previous dive was inherently dangerous because it involved a wreck that was tangled up in coral in the Atlantic Ocean. "And they were satisfied with that, so we made the dive."

Co-founded in 1971 by Link and John Seward Johnson, the son of the co-founder of Johnson & Johnson, Harbor Branch Oceanographic was first among peers in underwater marine science research. Among its notable accomplishments were the retrieval of the *Challenger* space shuttle booster rocket, which proved that the cause of the 1986 explosion was a faulty O-ring.[17] Farnquist's partnership with MacInnis and Harbor Branch burnished his bona fides further.

The expedition took place on July 3, 1994. Six trips were made to the wreck over three days. Harbor Branch's submersible vehicle, the *Clelia*, carried with it cameras and video recorders along with the same high-intensity lights used to view the *Titanic* in September of 1985.[18] The latter ship had sat thirteen thousand feet under icy waters of the North Atlantic Ocean, approximately twenty times the depth of the Fitzgerald wreck (530 feet). The combination of state-of-the-art photography technology and ideal lighting made the trip a success.

The Associated Press memorialized the moment when the world saw high-quality video of the ship for the first time:

> The *Edmund Fitzgerald*'s mangled carcass juts like a mountain from the flat, barren Lake Superior floor, rising as abruptly as the ship plunged to its death 19 years ago. The stricken giant's two main sections lie 180 feet apart, tons of taconite ore pellets from the cargo hold strewn between them on the gray silt bottom. Jagged rips in the hull, twisted steel plates and the bow's crumpled nose bear witness to the catastrophic force of the shipwreck. The windows of the pilot house are gone, blown away by the explosive rush of water.[19]

"It was quite the experience," Farnquist told me. By comparison, Fred Shannon was burnishing his bona fides as a gadfly. After Shannon dove to the wreck on July 25, 1994, three weeks following Farnquist, Shannon announced to the world that he found a body.[20] He then kited the idea of exhuming it and bringing the nineteen-year-old corpse to the surface. As well, according to Farnquist, Shannon was trying to convince one of the widows, Janice Armagost, that it was her husband. Why?

"He probably picked her because she was a weak link in the family members and tried to convince her, but she didn't fall for it," Farnquist told me when I asked why Janice. "She was the strongest one," Chris Farnquist added.[21] Chris had been coming and going into our conversations the last hour and into the living room, filling my coffee cup. "Yep, picked the wrong one," Tom said.

Both Chris and Tom attended Shannon's town hall press conference where he shared his findings from the trip and plans for releasing pictures of the body. If there was in fact a competition between Shannon and Farnquist, Tom was in the pole position. And that's right where he needed to be if he were to weather the coming storm of angry family members who weren't keen on the dives—Shannon's or his.

A recurring theme in my conversation with Tom Farnquist was the expansive professional network of diving experts, marine scientists, and shipwreck enthusiasts that he cobbled together over a career. He made this accomplishment seem effortless, as if he was organizing a pool tournament at a local bar or putting together a high school reunion. "So, the first time we dove the Fitzgerald, Ken Verancik worked for Michigan Sea Grant, and he started dabbling with ROV—remotely operated vehicles," Farnquist began. "He knew a guy out on the East Coast who had invented the ROV, Chris Nicholson. He worked with a good friend of mine, Emory Kristof from National Geographic."

This network made the difference. Without it, Farnquist could not have gotten his third expedition (1989, 1994, 1995) and first crewed effort off the ground. Everyone filled an important niche—photography, diving, research. But there was more. Not surprisingly his network of professionals had, in turn, additional networks of colleagues themselves who could assist the dive.

Emory Kristof, the famed photographer for *National Geographic* magazine, who specialized in undersea photography and snapped the iconic photo of the bow railing of the *Titanic*, had collected chits over time from technical experts who provided the camera and lighting equipment for his dives. Chris Nicholson made it possible for people to view underwater wreckage and rare marine life with the remotely operated vehicles he invented. It only took a few calls for Chris to secure investors who supported earlier dives. The actors were in place and the stage was set.

"After our dive in 1994, after Fred Shannon's dive in 1994, a spokesperson for the families called me, very upset, wanting to know why we were messing around the gravesite of their loved ones," Farnquist told me. His memory was a bit foggy, but he was certain it was either Ruth Hudson, mother of Deck Hand Bruce Hudson or Watchman Ransom Cundy's daughter Cheryl Rozman. According to Farnquist, Hudson and Rozman, along with Michael Argmagost's widow Janice, were key family leaders early on. Janice is the last living member of this trifecta.[22]

It was one thing for Oglebay Norton Corporation, Northwestern Mutual, and attorneys to take advantage of grieving widows, sons, and daughters and either settle quickly with insulting payouts or ignore the families altogether, as was the case with Mrs. Kalmon. It was quite another thing to desecrate the final resting place of their loved ones.

Since 1976, families and friends of the fallen sailors have gathered each year at the Maritime Cathedral in Detroit on the anniversary of the ship's sinking, November 10. On the anniversary in 1994, and with approval of the families, Reverand Richard W. Ingalls, Jr. invited Farnquist to the gathering to explain his actions. (At this point, the vast majority of families were beyond furious with Shannon and refused to even acknowledge him. His theatrics related to the discovery of the body were enough to put them over the edge.) "They [the families] invited me down to mariner's church to explain why we were diving and I said 'Well, for one thing we have a museum seventeen miles from the wreck site,'" Farnquist told me. "The first thing people ask us is about the *Edmund Fitzgerald*. And we had an exhibit on *Edmund Fitzgerald* because it was an important shipwreck."

> We [Farnquist and another member of the dive party] explained to the family members why we were there and the cautions that we took to be very respectful [of the bodies and human remains]. All we wanted to do was get footage that we could share with people because they're so interested in the Fitzgerald and Gordon Lightfoot as well. So, they were satisfied with that.

Farnquist has a disarming demeanor and soothing voice. He could put you to sleep with a lullaby in minutes or negotiate the release of a hostage. He was the diplomatic version of a five-tool utility player. It was a good skill to have when handling delicate matters like the Fitzgerald. But there was more. "How can we put this to rest?" the families asked Farnquist at the mariner's church about the whole ordeal, the punishing and tortured reality they lived for the last nineteen years. "Is there something on the ship we could bring up to serve as a centerpiece for a memorial at Whitefish Point?"

Farnquist recalled the exchange that would bring closure to many of the family members.

> "There's only one thing that I can think of aside from cutting half the pilot house off to get the name, *Edmund Fitzgerald*," Farnquist said. "'There's a 200-pound bronze bell sitting on the pilot house on top of three-legged belfry."
>
> "Well, how difficult would that be [to get it]?" the assembled family members asked.
>
> "It's not gonna be easy," Farnquist responded.
>
> "Would you help us?" they said.
>
> "It's in Canadian waters," Farnquist said. ("Some of them didn't know that it was in Canadian waters," Farnquist told me.) "Canada has claimed ownership of the riparian rights. And not only that, it's a kind of a national heirloom for them [Canada]. But, let me see what I can find out."

Thus began an improbable nine-month journey during which Farnquist and the family members cashed in chits, lobbied politicians, and pulled together enough equipment and expertise to dive the *Fitzgerald*. This was important not just because it would give families closure by marking a memorial with an artifact from the wreck. It was significant because of what it represented.

"It is the soul of the ship," Deborah Champeau, daughter of Third Engineer Oliver Champeau, told me, referring to the bell.[23] The bell was akin to the ship's personality or a birthmark, something that set it apart from other ships and could be easily recognizable and unique. Each bell rings differently, sounds different, and is used for different purposes. The *Edmund Fitzgerald*'s bell could be rung as it entered the Soo Locks, it could be used to greet well-wishers along the banks of the lakes, or it could be used to signal nearby ships. And it could chime all on its own. In bad weather, like pounding rainstorms or gale-force winds, the bell would peal wildly. The cacophonous ringing would betray the strength of the storm, something that was no doubt the case for the Fitzgerald in her last minutes. It was her final cry before she plunged to her death. Of all the things on the ship that could be taken and used for a memorial, the bell was the perfect choice.

Following a few calls with naysayers, Farnquist reached out to an old friend, a tried-and-true friend. "Emory Kristof and I started talking. I said to him, 'There's only one thing on that ship that's even possible to bring up and that is the bell.'" "Well, let's talk about that," Kristof replied.

The two settled into an informal yet very productive conversation. Kristof had dove thirteen thousand feet into the icy depths of the North Atlantic Ocean to visit the *Titanic* ten years earlier in the fall of 1985, where he captured unforgettable photos of the most famous shipwreck in world history.[24] He did so with a calm and grace that Farnquist could appreciate and emulate. It was a match made in shipwreck heaven.

Kristof's contributions would prove decisive in moving the ball downfield on the project. Just as Farnquist relied on friends and associates to make the 1994 dive to the Fitz, so he would also for this follow-on trip.

It wouldn't be cheap—$200,000 to be precise. But once again friends came through for him. Farnquist borrowed the sum from the Sault Tribe of Chippewa Indians. He had been instrumental in recovering the bodies of lost Sault fisherman over the years. He had a solid rapport with a tribal member, Dwight "Bucko" Teeple, who was more than happy to go to his tribal governing board to get the cash.

Now it was on to the hard part—international diplomacy. Because most of the wreck was in Canadian waters, they needed the assent of that government. While permission from Canada would be easier to get than, say, filling an extradition request with Mexico, it was still a tall order. First on the list to speak with was Farnquist's congressman, Bart Stupak (D-Michigan; 1993–2011). Stupak was a fierce advocate for the Fitzgerald families. He would take on Fred Shannon in 1995, calling him out for attempting to release footage of the body found at the wreckage site. Said Stupak on October 31, 1995, "I believe Mr. Shannon's intentions [to show footage of the body] are immoral and tasteless."[25] The former state trooper cum politician got to work quickly and got in touch with President Bill Clinton. For Stupak, Farnquist, and the Fitzgerald families, the stars were aligned.

Michigan was home to many Reagan Democrats, the archetype blue collar, conservative Democrat that floated into the GOP camp during the Reagan presidency in the 1980s but came back into the Democratic fold in 1992 when Clinton won the presidency. Reagan Democrats were a crucial constituency that Clinton had to keep if he were to win Michigan in 1996 in order to win reelection as president—a prospect that was in doubt after Democrats suffered a midterm rout in 1994 just a few days before the meeting at the Mariner's Cathedral.

That year, Democrats lost both congressional houses, and the president's approval rating slipped to 40 percent.[26] In just a few months, Clinton would argue to the Washington press corps that he was still "relevant" despite the GOP growing dominance in American politics.[27] Clinton was in the market for a quick win in a battleground state like Michigan. To get the bell, Clinton needed to secure the approval of Canadian Prime Minister Jean Chretien, a member of the Liberal party. Clinton requested the bell under humanitarian grounds.[28] It was a clever angle to win over a left-leaning prime minister predisposed to humanitarian arguments. Chretien soon approved, and the expedition was green-lighted. Farnquist and friends were in business.

At its core, the folklore of the Fitzgerald is about people. Everyday ordinary people. Throughout history there have not been many shipwrecks quite like the story of the Fitzgerald. It wasn't the *Titanic*, a luxury liner designed to carry the world's uber-rich across the Atlantic Ocean. Nor was it the *Andrea Gail* of *The Perfect Storm* fame that, while steered by a pair of blue-collar fishermen, meant little beyond the two and connoisseurs of fine ocean cuisine at upscale restaurants. The Fitzgerald was about hardworking men who prioritized the livelihoods of millions of other workers and families over their own safety. The Fitzgerald was the U.S. economy. It was up to crew members like Bruce Kalmon, Oliver Champeau, and Russell Haskell to transport taconite ore, the basic ingredient of the manufacturing economy, as well as steel, machine tools, paper machines, and tractors. Without these products the entire economy would come to a screeching halt.

According to Ric Mixter, by 1970, the one *billionth* ton of ore had shipped out of the Superior, Wisconsin, docks. That translated into roughly five hundred million cars. At the time the typical auto dealer was selling about one thousand vehicles per year.[29] There were literally millions of working families that relied on ore ships like the Fitzgerald so they could keep their jobs. In 1964, the Fitzgerald carried over one million tons of taconite through the Soo Locks, the first ore freighter to do so.[30]

Even if we were to find a foreign source of ore to make steel and would no longer need the mighty Great Lakes fleet—which of course we did—its loss would and did have cataclysmic repercussions throughout the economy. In one business quarter alone in 1982, the steel industry lost over 200,000 union steel jobs. One quarter. Dominoes tumbled from business to business, industry to industry, until manufacturing was a hollowed-out form of its original self. At the time of the sinking of the Fitzgerald over one in five Americans were employed in the manufacturing sector (22 percent). By the turn of the twenty-first century, it had fallen in half to one in ten (12 percent).[31]

Farnquist understood all of this, and he felt he owed it to the families to give them some semblance of peace and closure. It may have been high on his radar because of the economic pain of the Upper Peninsula in 1994. Michigan had experienced much tumult in manufacturing. The Wolverine State lost about seventy thousand jobs in the early 1990s.[32] Moreover, the Upper Peninsula was losing one of its major employers. KI Sawyer Air Force Base in nearby Marquette, Michigan, was recommended for closure in 1993 by the Base Realignment and Closure Commission and would shutter in 1995. The town of Sawyer's economy went into "steep decline" and twenty years later in 2016 nearly half (46 percent) of the population was living in poverty.[33] It may have been just one town in the Upper Peninsula, but it served as a cautionary tale for other communities facing economic uncertainty.

Farnquist was not about placating family members put off by his and Fred Shannon's expeditions the prior year. There was nothing political in his voice. He wasn't gaming out how family members would react. He didn't roll through a list of pros and cons. Nothing on his face betrayed any other motives than doing the right thing. After securing permission from the Canadian government and having already lined up the divers, funding, and equipment to make the expedition, Farnquist and his team were ready to go. Almost.

This dive would be nothing like the one in 1994. There was more involved than just looking at a sunken ship, taking pictures, and making notes. It would be quite an undertaking. They needed to remove a two-hundred-pound bell from the pilot house. Unlike the ship's keel that was left to rot over the years, the bell was in pristine shape. It was the soul of the ship and treated as such.

The bell was secured with three stanchions and in a way that would hold the bell in place under the worst possible conditions. This was especially the case for the Fitz, which was a rough-weather ship that never rode out a storm. "The Fitzgerald never backed away from a storm," Fitzgerald documentarian and historian Ric Mixter said.[34]

The way Farnquist told it, it felt like a spy-adventure film such as *Mission: Impossible*, in which Ethan Hunt must overcome not one, not two, but three rings of security at the CIA headquarters to get the NOC (non-official covert operatives) list of covert spies' real names. Or in *Raiders of the Lost Ark*, when Indiana Jones must dodge poisoned darts, miss wooden stakes shooting out of a wall, swing over a bottomless pit, and outrun a boulder to get a priceless statue.

To get the bell, Farnquist's team had an equally long to-do list and series of obstacles to overcome. But unlike Indiana Jones or Ethan Hunt, Farnquist was not an army of one. It would in fact take an army of divers, technicians, welders, military personnel, and photographers. First, they needed to put together the dive team. This was more than a tourist jaunt, which the Shannon dive resembled in comparison. They were not taking a bunch of Fitzgerald buffs to the ship to take pictures.

Meanwhile, Shannon and Farnquist's "rivalry" was coming to a rolling boil. Shannon sued Farnquist and the Shipwreck Society in May of 1995 to stop them from raising the bell. Was he truly interested in maintaining the integrity of the watery grave for future scientific research or was he bent out of shape that Farnquist was moving into the spotlight, and he was left out, sulking in the dark? The ruling judge answered that it was the latter. Ingham County (Lansing, Michigan) circuit court judge Lawrence Glazer ruled against Shannon on July 4, 1995, and effectively green-lighted the project.[35] The former officer of the law had gotten his ticket, and his points docked. Shannon soon faded away. Second, they needed transportation.

How would they get to the wreck and how would they move the equipment, particularly the submarines, into position?

Third was the dive dress. They couldn't take off the bell from a submarine. They needed a diver in person next to the ship to do the job. This also included finding the right diver with the right experience to make the dive and safely remove the bell and hoist it to the surface.

Fourth, the entire world would be breathing down their necks. Farnquist would need a team of advisors capable of handling the world press. This is to say nothing of all the politicians responsible for green-lighting the project who would bring to bear considerable scrutiny and perhaps blame if the dive fell apart. Farnquist was shouldering a load of expectations and work. It was daunting but not daunting enough to stop him.

As the old saying goes, it's not what you know but who you know. Farnquist began putting the team together with an early conversation with Emory Kristof. The basic requirement was the vessel to shepherd the party and equipment to the dive site. Farnquist needed a boat. Emory was a good starting point.

"Well, Emory says, 'I got a friend who's the admiral on the Canadian Navy,'" Farnquist recalled to me adding a slight chuckle, "It's kinda nice you know all of these important people." It was indeed nice. Kristof proved to be the key to making the 1995 dive happen. He was like the Kevin Bacon of the dive, except instead of six degrees of separation it was really two—at most.

It's not surprising that Emory knew all these people. He was a photographer for over sixty years, most of which were with *National Geographic*.[36] He was renowned worldwide for his one-of-a-kind photos that memorialized key moments in history. It would be like famed NFL coach Bill Belichick putting together a Super Bowl team or long-time Major League Baseball manager Dusty Baker doing likewise for an All-Star baseball team. Because of their long tenure as coaches, they knew a lot of people and, more importantly, knew which players fit in what slot and which ones worked well with each other. That combination is essential for putting together a top-notch team. Same went for the 1995 expedition.

Kristof began in earnest. He recruited Lieutenant Commander Robert Gwalchmai of the Canadian Navy, who happily obliged. Even more, Gwalchmai brought along two submarines to assist the dive, the Pisces 4 and SDL 1. Each carried four individuals—enough to do the job. Then there was the matter of getting to the ship to remove the bell. "Maybe we get one of the submarines down there to cut off the stanchion," Kristof suggested to Farnquist. "How the hell you gonna cut the stanchion off with a submarine?!" Farnquist responded. Kristof had the answer: Phil Nuytten..

Nuytten was the inventor of the Newtsuit.[37] It was a modern day, state-of-the-art dive suit that could handle deep-water projects. It offered the diver the comfort and protection necessary to not just explore ships but, in the case of Farnquist's dive, remove a piece of the ship with a blowtorch.

The Newtsuit was flown in from Newton's home in British Columbia. Kristof brought in high-definition video equipment courtesy of Sony Corporation. Famed movie director James Cameron lent his U/W lighting (i.e., underwater), the same technology and equipment he would use to film *Titanic* the following summer in July 1996.[38]

It was not just nuts and bolts. There was more to it. They were putting on a show for the benefit of the families. Yes, the objective was to retrieve the bell to serve as a memorial for the families, but it was also an experience. This moment would be among the most significant and memorable of the families' lives. And it had to be done just right. "We choreographed it all first," Farnquist told me.

Many of the family members would be awaiting the bell's removal. They would be the first to glimpse it as it was pulled out of the water. The real drama was underwater as Bruce Fuoco in the Newtsuit carefully cut the bell off the stanchions. It was almost not to be. Farnquist, a raconteur of the first order, retold the story as if it had happened yesterday.

"Everything went perfect. Except. We're talking to the guy in the Newtsuit and he's saying, 'Come on, baby, come on, baby.' He's trying to burn that last leg off from the ship. And finally, it broke loose, and that stanchion broke this close to his face shield," Farnquist said, holding up his finger and thumb just two inches apart. "If it would have hit his face shield, we would have 30 guys down there on the Fitzgerald." The stanchion would have punctured the diver's mask and he would have died instantly. "That was a close call," Farnquist said.

The bell, attached to a chain, was carefully pulled to the top. As it reached the top and broke the surface, it did something unexpected. "When the bell came out of the water it actually swung on the cable and tolled—bong, bong," Farnquist imitated for me. "So, there was a lot of emotion going on there," Farnquist said.

Three days later on July 7th, the families gathered together on a boat directly above the Fitzgerald. Each family member rang the bell and dropped a carnation into the lake. James Haskell, brother of Russel Haskell, was there. "That's the closest I was gonna get to my brother," Haskell recalled to me from his home just outside Sturgeon Bay, Wisconsin. "What was really nice about it, after we went through all this stuff, is that when we came back home it was something my cousin and her daughter and myself could talk about."[39] Jack Champeau, son of Oliver Champeau, said the bell recovery was "very emotional" and would hopefully finally put it to rest. "I believe that this will finally put this to an end and we can rest now as I'm sure he [Oliver Champeau] is," Jack said on the occasion of the bell retrieval.[40]

Ironically, the weather was far from pleasant. A contemporary account of the day noted its eerie resemblance to the day the Fitzgerald sank. "Friday's unseasonably cold weather, with dark clouds and fierce wind gusts, was reminiscent of the day the Fitzgerald [sank]."[41] "It was so cold," Farnquist told me.[42] On the way back to the Soo Locks they happened to pass a boat that was upbound—the *Arthur M. Anderson*. The Anderson was the first of three ships that went back out into the storm to try to rescue the Fitzgerald and her crew. The odds of the ship passing the families was, according to Farnquist, "next to impossible." But it happened. "It was a moment described by the families as an 'omen of final closure' as did everyone on that special day."[43] And a well-deserved one, too.

CHAPTER 10

Hard Knock Life

I GREW UP IN A BLUE-COLLAR NEIGHBORHOOD, CAROL LYNN DRIVE IN LITTLE CHUTE, Wisconsin. And on that block all the dads and some moms worked at a paper mill, except for my dad, who was a Lutheran pastor.

I learned two important life lessons growing up. First, the importance of serving your neighbor, community, or congregation, to do what you can to make your corner of the world a better place to live. Second, there are a lot of people who work hard, play by the rules, pay their taxes, but cannot get ahead, because there is too much power and too much money concentrated in too few hands.

We were a far cry from Wall Street. So too was Iron River, Wisconsin in the far northwest corner of the state. But the difference was we were home to the top paper-making economy in the country with no shortage of good paying, family supporting jobs whether you had a college degree or a high school diploma. The mills generated a mega economic ecosystem worth billions of dollars that created white- and blue-collar jobs alike, to say nothing of the countless main street diners and coffee shops, medical offices, machine tool factories and schools that depended on the mills to support their business. Iron River and most of the Northwoods had limited job opportunities, and the work that was available was hard and back breaking: logging, farming, construction labor, and longshoring.

Taking a few detours over graveled roads just off County Road H in Bayfield County, a few miles south of Iron River, roads so worn that pocks of clayed dirt showed through the roadbed, I came across a modest, one-room cabin. Walking up to the front door, I was reminded of one of those historical settlement houses at a state park, perhaps a lodge for a fur trader. Cord wood was stacked along one side of the house, and a chimney poked out of the top of the roof's pitch. Smoke was billowing out despite it being an early summer morning with suffocating humidity. Perhaps there were no gas hookups or even electricity.

A gentleman came to the door. He shot me a quizzical look as if to say, "Who the hell are you?" I must have looked like the county surveyor or even the tax assessor. I didn't see a side arm on him nor a rifle laying up against the door jam. I should be okay, I thought. "I'm looking for Fred Armagost," I said. "That's me," the man answered.

Fred was the little brother of Third Mate Michael Armagost. The years had not been easy on him. He looked tired and worn and walked with a small gait as if he was carrying an extra load. It could have been alcoholism—a disease prevalent in one of the top states in adult alcohol consumption.[1] It could have been the loss of his older brother at a young age and at a time when he could have used a big brother. It could have been a long life of hard work and few breaks. "I've logged for many years. Cut a lot of trees," Fred told me. "It's dangerous work. I got lucky a few times. Every day you get up. Keep your eyes open, pay attention or otherwise it can happen real quick."[2] His sentences came out quick and with a staccato beat. He was a fan of word economy, said only as much as was needed to get his point across.

Fred handed me a lawn chair to sit in near his front stoop. I had to navigate around an extended ladder that reached the top of the house, careful not to knock it over or walk under it for fear of bad luck. But I had found a close relative of a fallen sailor of the Fitzgerald, so it was my lucky day.

It was also my good fortune because incredibly, after forty-seven years no one had bothered to interview Fred. I was the first to tell Fred's story. "You must get these interview requests all the time," I said. "No. You're the first," he responded.

The insights would either affirm my initial impressions of the Fitzgerald families—tough, resilient, patient—or shed new light. I got a bit of both. "You lose the ones you love," he told me. Mike wasn't the only family member he lost. He lost his father when he died of a heart attack, age forty-seven. This was a recurring theme among the survivors: losing loved ones at a young age.

Fred's words trailed off. He had a low drone to his voice and a sloppy lisp that was hard to make out at times. I wanted to hear every word. He was living history, the brother of a Fitzgerald crew member, a descendent of Midwest folklore. I leaned in, though careful not to collapse the lawn chair on my back.

Fred remembered the very moment when he first learned the Fitzgerald went missing, or "went off the radar," as he recounted to me. It was a crystal-clear recollection that was impressive, considering "he was poor at names" and his memory wasn't good. Or so he said. "I was staying with my sister and my nephew. And I'm watching the news. And it [the missing Fitzgerald] came on the news. And they said it wasn't on the radar. I knew right away it was gone. No doubt," he told me. "So, I walked over to my [INAUDIBLE] house. Which is just down the road aways. And my sister-in-law is there and my mother and my niece and nephew and I come in and they looked at me and I just simply—why have hope when there is no hope."

Michael's widow, Janice Armagost had a similar account. "I started shaking and I just knew. Mike had always said that the water is so cold in Lake Superior in the fall that if you ever fall overboard, you may as well forget it. And I just knew it was all over at that minute," she said.[3]

Fred Armagost had always been a family man. He knew what was important in life. Yes, paychecks put food on the table and paid doctor bills; but something else was much more important. "I had four young kids, so I quit sailing," Fred told me. Fred took after his older brother and sailed the lakes for years. You could say his family-first credo added up to good survival instincts. The last ship he was on was the Fitzgerald and left his job shortly before it sank in early 1975. Fred never went to the ceremonies or reunions with the Fitzgerald families. It was too personal for him. "I didn't feel great going," he said. Could you blame him?

Janice never remarried, and there isn't a day that goes by that she doesn't think of Michael. "I look at my kids and I see Michael," Janice said. Her personal memories and family are not the only things that remind her of the fate her late husband suffered. It doesn't help when Gordon Lightfoot's song comes on the radio, or someone recognizes the last name and asks about her late husband. Nor does it help that every well-wisher invariably follows up with the question, "So why do you think it sank?" "Nobody realizes that there are survivors," Janice said. "I mean, my kids' father is on that ship and my husband's on that ship. And people just think of it as a shipwreck that happened so long ago, and it's not."

When a group of Fitzgerald buffs led by Michael Shannon went down to the Fitz in the summer of 1994 and a documentary was released shortly after, Janice was forced to relive everything. It wasn't just the knowledge that strangers were poking around her husband's grave site. "I was always able to stop myself from thinking up to a certain point, when I think of Mike on the boat," Janice said. "As soon as I would reach a certain point like the last three, four or five minutes of his life, I was always able to just block that out of my mind and never really come to grips with it. But now this is making me deal with more things. And that, of course, makes you think what went on in somebody's mind the last five minutes, the last 10 minutes before they died."

Every Fitzgerald family member I met handled their loss in their own way. This is not surprising. Everyone remembered their relative well despite their many flaws. An estranged father. A stepdaughter who had no relationship with her stepfather. A soon-to-be divorced husband. And what everyone wanted more than anything else was the approval of their lost father or son or husband, pride in what each of them became no matter their lot in life.

"And you know, I just got really good at what I was doing," John O'Brien, son of Wheelsman Eugene "Red" O'Brien told me. "That's one thing I do wish my dad was around because he always thought I had such good morals."[4] Today John is a multi-millionaire in Fort Lauderdale, Florida, fueled by successful ventures and investments. And like Fred Armagost, he inherited the sailing gene and keeps a slip at the nearby harbor and sails whenever he can.

Deborah Champeau, daughter of Third Engineer Oliver "Buck" Champeau, devoted her life to her father's memory, taking all of his many life lessons to heart in the hope that he would have approved. "There's been points in my life where I missed him really bad because he didn't see me get married. He didn't see my children. I could feel his absence in my life. And I always wanted him to be proud of me," Champeau told me.[5]

If you could drop a pin on a spot on a map of Wisconsin that was furthest away from Iron River and which was also on the opposite end of a demographic or socioeconomic chart for wealth or general health outcomes, it would be Nashotah, Wisconsin.

Rolling hills and meandering roads run through countless subdivisions in this exurb of Milwaukee, thirty miles from its downtown. Among the neighborhoods of Nashotah are quaint main streets that remind a visitor of Martha's Vineyard or Mackinac Island.

Each subdivision is anchored by a cul-de-sac behind which sits an expansive corn or hay field to remind you that you don't live in the city. There is little that unites this far-flung suburb with Milwaukee. The region is one of the most segregated metropolitan areas in the country.[6] It is home to the state's poorest zip code in downtown Milwaukee as well as the wealthiest, Okauchee, an exurb that sits due east on the very edge of the metro area.[7]

Nashotah is home to Deborah Champeau, daughter of third assistant engineer, Oliver "Buck" Champeau. Her journey to this bucolic corner of the Dairy State could only happen in America. Her father came from humble beginnings. He was forced to drop out of school when he was a sophomore in high school to provide for his family, just as Armagost did for his brothers and sisters after their dad died young. They grew up in a modest neighborhood in Milwaukee.

Oliver Champeau began his career on the lakes as a merchant marine, which took him all over the world. He never forgot to bring home a gift or two for his only daughter. "He went all over the world. I have dolls from everywhere or souvenirs and postcards from countries like Turkey and Mexico. I mean just all over the world," Champeau told me.

Later Oliver went back to school and earned his high school diploma and degree in engineering to work on an iron ore ship on the Great Lakes. Champeau noted that while her father was qualified to work as the second engineer, he took the position of third engineer because it was the only post available. Champeau wanted to be on the Fitzgerald above all else. "He was honored to be on it, because it was so well known in the American Great Lakes fleet," Deborah said.[8]

Going back to school positioned Oliver Champeau to rise through the ranks until he was an engineer on the *Edmund Fitzgerald*. But that wasn't all it accomplished. He set an example that would guide Deborah through her life. Said Deborah, "He just was a very strong man for having to quit school sophomore year to take care of his brothers and his sisters and his mother. Sending checks home. He was in the Marine Corps. Just supporting his family and that was who he was. If he could help you, he'd give you the shirt off his back. So, when people tell me I go overboard trying to help people, well, not so when you have an example like that."[9]

Oliver was very protective of Deborah and handled her like a porcelain doll. Every privilege she was granted—a new car, Saturday night out with friends, allowance money—came with a set of rules. It seemed to work. She still carries "his list of rules" with her today. He was strict, but he wasn't a disciplinarian. "I can't say that he ever yelled at me. I can't say he ever punished me. He just wasn't like that," Deborah told me.

He was gone nine months out of the year and spent a good chunk of his paycheck on long-distance phone calls. He had to be there for her in some way; especially for life's big events, including the biggest event for a high school girl:

junior prom. "I got asked to prom and I didn't know what I should do," she told me with excitement and a wisp of nostalgia. "I didn't know how to get ahold of him. I mean, the phones were not like how they are now. And so, I wrote this long letter to him about how much the dress was at Boston Store, how much my shoes were going to be, and how my hair had to be. My grandmother—his mother—got the letter to him. I don't know how she did it, but she did. And he called me up from a port and said, 'Okay, you've got the exact amount—now go get your dress, your shoes, and get your hair done!'"

It wasn't too long after her prom that her father perished along with the other crew members. But she refused to believe he had gone. She couldn't imagine life without him. It would take Deborah many years, until her wedding day, to finally understand her father would never come home. "I don't think I really believed he was gone until my wedding day," Champeau told me. "I thought, you know, he'll be here, even though—but no, he's not walking me down the aisle."

Anyone who loses a loved one unexpectedly carries some regret. Regret they did not spend enough time with them. Regret that they never said I love you before they died. Regret for a silly argument they had with them. Some people blame themselves in part for their loved one's death. Deborah was one of them.

Like many Fitzgerald sailors who come from humble backgrounds, seeking out opportunities to make additional money was second nature for Oliver. For those willing to work longer and harder and take risks, there was no shortage of hazard-pay jobs. Deborah felt responsible for the pressure that her father was under to make more money to pay for family essentials—clothes, mortgage, food, school supplies—but also the extras, like a cabin up north in Door County, the peninsula or "thumb" of Wisconsin. "In November it's so unseaworthy that you get extra pay for being out there," Deborah told me as the timbre of her voice changed, searching for the memory itself in the recesses of her mind. It was hard for her to talk about something so personal. "My dad thought he was going to pay off his retirement home up north that way. That really affected me because I was part of it—what color room do you want? What color carpeting?" The house had been finished that fall.[10]

Parents take on hard work and jobs they don't want to because that is what needs to be done. The life of a family in a steel town could be punishing, as Gabriel Winant outlines in his book *The Next Shift: The Fall of Industry and the Rise of Health Care in Rust Belt America*. Work was demanding. A spouse would often augment the salary by pulling a night shift, but that is what needed to be done to put food on the table, a roof over their heads, a solid education. Or Jack Metzger's memoir *Striking Steel: Solidarity Remembered,* where he accounts a tough but nurturing life that provided ample opportunity for him and his contemporaries. "I and my generation were (almost) all living well—not only materially, but mostly with warmth, commitment to each other, and a good deal of integrity," Metzger wrote. "None of us had to kiss anybody's ass or to screw over people to get ahead."[11]

Every generation wants to leave the world a little better for the next, to make sure their kids have a shot at the American dream even if they themselves missed their opportunity. It is a fundamental element of the social contract, something Winant and Metzgar understand and every son or daughter of a Fitzgerald sailor lived.

The late Carrol Quigley of Georgetown University, renowned professor of western civilization, shared the concept of "future preference," which dovetails with the idea of the social contract. He taught this lesson to his freshman students in an effort to explain how they found themselves in his classroom. "The thing that got you into this classroom today is belief in the future, belief that the future can be better than the present and that people will and should sacrifice in the present to get to that better future." Quigley went deeper.

> That belief has taken man out of the chaos and deprivation that most human beings toiled in for most of history to the point where we are today. One thing that will kill our civilization and way of life is when people no longer have the will to undergo the pain required to prefer the future to the present. That's what got your parents to pay this expensive tuition. That's what got us through two wars and the Depression. Future preference; don't ever forget that.

The argument was so compelling, one student and future president—Bill Clinton—would weave the theme in nearly every campaign speech he delivered.[12] Whether Clinton heeded his mentor's advice and reflected Quigley's worldview in his policies as governor and later president was another matter. No doubt each student took a different lesson from Quigley's teachings. Although he most undoubtedly meant well, not all truly learned this lesson or, as Jon Shelton writes in *The Education Myth*, "Clinton pushed education as a magical panacea that could, by itself, solve the nation's growing economic inequality and serve as the answer to the threats capital flight and globalization posed to blue-collar livelihoods."[13]

What applied to Quigley's students applied to the sons and daughters of the *Edmund Fitzgerald*. Deborah's father made her life in a well-to-do suburb possible; she would do all she could to memorialize her father's legacy and make sure the future was full of opportunity and possibility for his grandkids.

Deborah was one of the more outspoken family members and played an integral role in lifting the ship's bell from the wreckage in 1995 as well as helping to pass a Canadian law that forbade shipwreck explorers from disturbing the site. It's an accomplishment she has been quite proud of. "I made sure you couldn't dive down there. I was instrumental in getting the bell. I was instrumental with Canada. I was instrumental with protecting and I will be that way for the rest of my life," she proudly told me.

The bell was more than just a part of the ship, Champeau told me. "It is the soul of the ship." The bell could be heard over the deck and throughout the ship and beyond. It served as the ship's personality. Among the ships, each bell rang and sounded differently and could be identified accordingly. In that way, the Fitzgerald's bell personified her father. "He was jolly he had a laugh that filled a room, and he always had a smile," she said. People recognized Champeau for his unique traits and character, just like the Fitzgerald.

In Deborah's waking hours and at night, her father's memory is always with her. He'll always be the protective father looking out for his daughter on prom night, getting a bible for her first communion, or making sure she had a full Easter basket. "It haunts me. I have this dream. I'm swimming and I see a hand come out of the water. I'm trying to grab it, to save him, but I can't grab it. I would wake up and I'd be sweating," Champeau told me, struggling to keep her thoughts in order, a stream of memories, emotions, and regrets all coming together at once. "Knowing there was nothing he could do. The lifeboats, knowing what they look like and how traumatically ripped apart things were. I wondered what he exactly felt and went through and what was he thinking, and did he know that the inevitable was going to happen?" She'll never know. And that's what Deborah carries through life.

John O'Brien has been a successful entrepreneur in Fort Lauderdale, Florida. He owns and runs twenty-five laundromats and distribution centers in addition to other profitable ventures. John was the good son, and not just because he was Wheelsman

Eugene "Red" O'Brien's only child. He may have moved away from his native Ohio, but he never forgot where he came from. He is reminded of it every day he goes to work. I was intrigued as to why he went into the coin laundry business. It was a far cry from the family business; to say nothing of landing in the Sunshine State, whose weather and way of life is the opposite of hardscrabble, hard weather Ohio.

In his early twenties, shortly after his father passed, John answered a sales ad on TV that pitched a business: Be the owner/operator of a laundromat and make $80,000 per year! "It cost about $200,000," John told me. With the savings that his father had given him over the years and the insurance settlement from the sinking (about $25,000), he made a down payment and took out a loan for the balance. "I've been making money in the business ever since."[14]

Like Deborah Champeau, John has put his father at the center of his life and has credited him for the man he became. "I'd like to think my dad would be proud," John said. "I take pride in being responsible and doing the right thing. I got that value from my father." Every child tries to impress their parents, and they show it in different ways. For Deborah, it was carrying simple yet important lessons to guide her through life. And for John it was success and riches in business. By the measure of their accomplishments, they each certainly would have met if not exceeded their father's expectations.

John has incredible demands on his time and works hard just to find more time in the day to do business. "You want to meet for breakfast at 6? I'll beat you to breakfast and you better be there by 5:45 because the last one pays." I was beyond grateful that he found not one but three appointments to speak with me (and it didn't cost me three breakfasts).

John has a healthy self-awareness that would put any psychiatrist out of business. According to John, he has made more money in a month today than what his first house cost. "I'm not bragging, I'm just saying." He doesn't have a southern accent, and I am sure he lost his Midwest inflection a long time ago. His voice reminds me of a mafioso, but at the same time it's warm, caring. He comes across as genuinely interested in you, like a father would be toward a favorite son. I think his voice also reflects the confidence and command of business that he demonstrates more than anything nefarious. I can't see him breaking a single law or crossing a business partner, and for a simple reason: his father would not approve.

John understands the social contract and Quigley's concept of future preference. Without a dad, he has had to rely on mentors to help him in business and guide him through life. They are part of his success, he insists. Now he pays it forward, seeking out young people, "hustlers" and "good family people," as he puts it. And he does so not just because it is the right thing to do but because it makes good business sense. "If I bring in one them as a partner, he's going to work ten times harder than me. He'll make at least the same amount of money I would," John explained to me. "But in this case, I'm putting someone to work, and they're going to remember me and help me make even more money down the line."

His dad wasn't perfect—whose parent is? The work was hard on family life and took a toll. His parents' marriage ended in divorce when he was young. He explained to me no fewer than four times that the job of a Great Lakes sailor was not a good family job. "You're on the lakes for nine months out of the year and only in port for eight hours. Most guys did not live in the home port. They can't drive home, maybe the wife and kids meet them at the boat maybe you'd see families meet people at the boats. It's not a really great family job," he told me.

But it was clear Red O'Brien had no greater priority than his son. It was the same instinct that governed Oliver Champeau's life with his daughter. For Red his love for his son came in the form of a nice nest egg from his weekly paychecks to give John a good head start in life as well as maximum father-son time when he was in port. That was not all the time he would set aside for his son. "He liked to write letters. And he wrote a letter to me once a month. Whenever he was in town, I would always make sure I was around to see him when his boat came in every time. He'd picked [*sic*] me up and we'd hang out for the day. He was a large part of my life," John told me.

To get to James Haskell's home in rural Sturgeon Bay, Wisconsin, you must wind through a country subdivision not unlike Deborah Champeau's part of Nashota until you pop out off an inclined street and onto Taube Road. Haskell's modest ranch house, set back a good acre from the road, contrasts with the up-to-date homes in its backyard, like modernity closing in on yesteryear.

Haskell and his mix husky-corgi meet me in the driveway. He shows me around his home, particularly his Fitzgerald memorabilia. Inside his house, I negotiate around a La-Z-Boy and stacks of old magazines and VHS tapes, along with a few end tables piled with empty cereal bowls, mail, and an ash tray. My iPhone can't get reception, and my recording app won't work. We go outside to get a better signal for the interview.

James reminds me of Fred Armagost. He could pass as his brother. The first thing that crossed my mind was that he was someone who had led a hard life. Next, I sensed that deep inside he longed to share with others his melancholy that had consumed him for the better part of fifty years.

He was open about his history with the Fitzgerald. His face beamed when he recalled his trip to Whitefish Point for the twentieth anniversary of the sinking in 1995, when the bell was brought up and each family member rang a toll. Russell's widow, Evelyn, would not make the ceremony. She had died three months earlier.

Even though he had not seen Russell since Christmas of 1970, five years before the Fitzgerald sinking, the memories of his dear brother are still fresh. I found myself in cell range in the middle of James's front yard. That was the good news. The bad news was that I was on top of an ant hill. Good enough.

The distant laboring sounds of hardworking tractors pulling a second harvest off hay fields could be heard as well as a car or two rumbling down the road. Somewhere between Orphan Annie and Jay-Z lies the truth; small-town America suffers a "hard knock life," feeling the hurt and struggles against impossible odds every day to make a living, put food on the table, and rise to fight another day. It is not just material for a Broadway hit or a rap song.

Said one reviewer in 2011 of Jay-Z's hip-hop version of "Hard Knock Life," comparing it to the original song, "[The songs] may [each] seem diametrically opposed, but they end up being fundamentally about the same thing: forming a common thread of struggles with poverty spun into gold."[15] For most families of the Fitzgerald, no great riches would come from their struggle, nor would they make millions of dollars in royalties off their hardships.

From a distance, James Haskell watched his brother suffer in ways no human should. In the summer of 1975, Russell Haskell lost a daughter. Russell and James had lost their mother just two years earlier in 1973. "The youngest girl got hit by a hit and run driver and was killed. Three months later the ship went down," James told me.[16] And just as the sinking of the *Edmund Fitzgerald* remains a mystery to this day, so too does his daughter's assailant.

Russell had plied the waters of the mighty Great Lakes since age nineteen. He was known for his good humor and bonhomie. After his youngest daughter was killed, though, he fell into a deep depression. It didn't help that Russell had survived his own car accident. When he was eighteen years old, he was thrown from a friend's car and fell into a coma for thirteen days. He missed his high school graduation.

"It was awful. His wife lost her daughter and a husband in three months," James said. Like Armagost's widow Janice, Russell's widow never remarried. As I struggled to make sense of Haskell's misfortune, a couple from next door walked past on the shoulder of the road and said hello. It was a relief from the growing sadness of our conversation. I could tell that James appreciated the break. James's dog had a remarkable sense of self-control and self-awareness. Even though the next-door couple had a dog themselves, James's hound just hummed a low growl and didn't run after the chocolate lab. He wasn't on a leash and James certainly did not have invisible fencing. "He's not going to run after them?" I asked James. "You know what he would do?" James said. "He'd get up run back through the door, into the cage and then bark!" he said with an approving smile. We weren't treated to a unique canine temperament of calm and restraint, but it elicited a good laugh anyhow.

James's highpoint in life came in the mid- to late 1960s when he was in the U.S. Army and serving overseas. "I got ready for the draft, so I enlisted in the Army [in 1965]," James told me. "They sent me to Japan, [to the] biggest [radio] relay station in the Far East." In 1968, he had a couple-day shore leave and decided to climb Mount Fuji with a friend. "We started at three o'clock in the afternoon and climbed until 8:00 p.m. We went to sleep, opposite like sardines when you pack them in," James told me with a soft chuckle. "We get up at two o'clock in the morning and climbed [down]. It was sunrise at the top of Fuji. And then you shuffle your feet on down."

It was 1965 and the Vietnam War was a few years from coming to a rolling boil. He was twenty years old, and life was full of possibilities. To think enlisting in the Army in the early days of the Vietnam War was a fond memory and life's high point for him really put into perspective the rest of a life that was filled with loss, back-breaking work, and disappointment.

Like Fred Armagost, Deborah Champeau, and John O'Brien, James Haskell can remember the moment he found out that the *Edmund Fitzgerald* went down. "The news of the ship going down was on the morning news. My sister heard it first," James told me. He was busy at work at a lumberyard in Sturgeon Bay. James got the news around the same time Fred Armagost did up in Iron River. "[My sister] called my brother, asked him what ship that Russell's on," James said. "Then they called me all day long with updates." Despite the crushing news and knowledge that his brother was lost at sea and presumed dead, James kept on working. The lumberyard was down to only two drivers. James had to watch the front office and "take care of the customers." That's what needed to be done. James dutifully reported to work the next several days until the loss of the ship and all souls aboard was finally confirmed.

"I think my father would have been really honored to know that the men he was with, his brothers, that they're all remembered November 10th," Deborah Champeau told me as we were wrapping up our nearly one and a half-hour long conversation. It's easy to see how the family members of the sailors of the Fitzgerald considered each other family as well. They sought to accomplish important goals in the name of their relatives, particularly the raising of the bell and getting the Canadian government to declare the wreckage site off limits to divers and others who might disturb the final resting place of their fathers, brothers, husbands, and sons.

The wreck of the *Edmund Fitzgerald* affects more than just immediate family members and friends. Many understand this, including Deborah, thanks in no small part to Lightfoot's powerful ballad that generated worldwide interest the year after the sinking in 1976. Family members see a role they must play, an obligation they must fulfill to keep the memories of their family members alive. When you speak with Deborah, it sounds a lot like someone who lived through 9/11.

"I've talked to many people to see if there is a family member or how they're affected by November 10," Champeau said. Other family members play down its significance and think their father or husband would be nonplussed. "My dad was a low-key guy," John O'Brien told me. "And I don't think all this would ever matter to him. He was never trying to be anything other than who he was."

There are only a handful of "first-generation" Fitzgerald family members left, a few wives and siblings. There are more second-generation members—sons, daughters, nieces, and nephews. There is even a third generation that understands the importance of the Fitz's legacy, as well as that of fulfilling a familial obligation. "I know they'll carry it on. I was just talking with my eleven-year-old grandson, and he told me I have nothing to worry about," Champeau said.

At a bell-ringing ceremony on a recent anniversary in Whitefish Point, her grandson gave a moving tribute to his great-grandfather. "He was only six years old. And he wore a sailor's hat. And he walked up to the bell on the deck with my son. He rang the bell and then he stood back and saluted. He saluted it perfectly. And I just about lost it," Champeau told me. In the meantime, Deborah is at the ready to protect the legacy of her father and the ship itself. Those who dare cross the likes of Deborah or other family members, you've been warned. "I may not have any power at all," Deborah told me. "But if they get my Irish up in me—it's a different story."

CHAPTER 11

Great Nations Build Things

EVEN THOSE WHO DID NOT WORK IN PAPER MILLS ON MY BLOCK OWED THEIR livelihoods in part if not entirely to the industry. Public school teachers and county highway workers depended on a strong manufacturing taxbase to fund their salaries and benefits. Supper clubs were built around Friday night fish fries, the same day that many patrons received their paychecks. Hundreds of small businesses lining main streets, packing business parks, and filling strip malls lived or died on a strong manufacturing economy that pulled in billions of dollars of economic activity every year.

Little Chute, Wisconsin, and the surrounding Fox River Valley of northeast Wisconsin boasted the highest concentration of paper mills and plants in the world. The region was nicknamed the Paper Valley. While it still holds that moniker today, the scope and scale of its footprint, economically and physically, is a much smaller version of itself.

Relative to other manufacturing industries, the paper industry was a lagging indicator of the sector's troubles. In 1990, northeast Wisconsin held 15,896 pulp and paper industry jobs. But at the end of the century, it had dropped to 14,859, a 6 percent decline. And by 2010, three years into the Great Recession, it had declined precipitously by 23 percent to 11,405 jobs.[1] Driving through Kimberly, Wisconsin, opposite Little Chute on the Lower Fox River in 2010, as a 119-year-old mill was shuttered, I felt like Michael Moore as he drove through his economically depressed hometown of Flint, Michigan, in his 1989 documentary *Roger and Me*. If GM was viewed as the villain in Moore's film, then the antagonist in Kimberly's dramatis was NewPage—but with one difference: NewPage was owned by Cerberus Capital, one of the largest private equity firms on Wall Street. To Cerberus partners, the Kimberly NewPage mill was just another line item in their expansive, worldwide portfolio.

Kimberly's next-door neighbor to the east, Combined Locks, would dodge a bullet later that decade when its mill, Appleton Coated, went into receivership.[2] Menasha's Whiting Mill, U.S. Paper Convertors of Grand Chute, Neenah Paper of Appleton, and Clearwater Paper of Neenah were not so lucky. They took hundreds of years of paper-making traditions and well over a thousand jobs with them when they closed in 2016, 2017, 2021, and 2021, respectively.

"Great nations build things."[3] That's what the CEO of Minnesota-based semiconductor producer Skywater Technologies said when the U.S. Congress passed the CHIPS and Science Act of 2022. The legislation allocated $280 billion in new funding to rebuild the country's semiconductor industry. In 2022, the United States produced just 12 percent of the world's semiconductors. Asia is responsible for making 80 percent of the world's microchips.[4] It is safe to say that U.S. consumers utilize more semiconductors in the products they use, from refrigerators and toasters to cars and motorcycles, than any others in the world. It is not a sustainable trajectory. The United States learned its lesson in spades during the COVID-19 pandemic, when household durables had yearlong back orders. This was the impetus for the legislation.

Unlike policymakers in the wake of the COVID-19 crisis, the American conscience wasn't shocked into action to rebuild the U.S. shipbuilding industry when it hit its nadir over forty years ago. Then again, it is easier to make the connection between household durable goods and microchips. Every contraption necessary for modern life depends on plentiful access to semiconductors. Microchips and jobs were also embedded in the minds of policymakers who recalled former presidential candidate Ross Perot's sort-of catchy phrase from the 1992 election year, "You make more making microchips than potato chips."

While semiconductor manufacturing promised good jobs and a strong economy, the pivotal role of the shipping industry wasn't as obvious. Shipbuilding conjured up images of foreign competition or sending jobs overseas. Those Great Lakes ore ships like the Fitzgerald sure look a lot like massive container ships moving crates chock-full of cheap imports that flood markets and kill good jobs. As the shipbuilding industry withered on the vine, the response by Americans and the politicians seemed to be, "Maybe it's just as well."

The United States has a troubled history in shoring up key industries that are responsible for creating millions of good-paying, family-supporting jobs. There is no shortage of contemporary examples of the government offering significant preferential treatment to select industries. The banking and financial services industry is an obvious example. The Troubled Assets Relief Program of 2008 allocated a half-trillion ($456 billion) to buy up bank stocks and mortgage-backed securities.[5] Not only was the fate of the U.S. economy hanging in the balance but so too were election prospects. In the coming election cycle, the finance and real estate industries would be the top GOP donor and the second-largest donor to Democrats.[6]

Among the other industries hurt in the Great Recession, it was the auto industry that was singled out for substantial government assistance, to the tune of an $80 billion bailout.[7] This was a curious choice, because unlike the early years of the Fitzgerald where the auto industry dominated, the auto industry share of GDP on the eve of the Great Recession (2007) was just 60 percent of what it had been prior to the 1973 oil shock.[8] Yet there was history to this decision. The United States government had granted Chrysler $1.5 billion in guaranteed loans in 1979; it made $300 million on the deal,[9] just as it would make money after the 2009 auto bailout. Also, the American auto industry remained a cornerstone of the U.S. manufacturing economy at least symbolically, which counts a lot in politics and the vast majority of jobs at the Big Three automakers (GM, Chrysler, Ford) are organized and unions are a key component of the Democratic Party, which held all levers of power in Washington at the time.

If you go deeper into U.S. industrial history, you see ample examples of blatant favoritism. The railroad industry in the nineteenth and early twentieth centuries come to mind. Like the banking and financial services industry of today, the railroad industry had politicians in their front and back pockets. AMC's *Hell on Wheels* (2011–2016)[10] was a historically fictional TV series but was not far off the mark with its storylines about Congressional committees approving (and disapproving) the sale of land to the show's subject, the westward-expanding railroad giant Credit Mobilier. One of the program's central plots hued closely to the real-world fraud case scandal that occurred in the immediate post–Civil War era.

Even when workers reaped some benefits, the intentions of industry titans or lawmakers were not good, to say the least. One notable example was the Smoot-Hawley Act of 1930. It was intended to shore up an industry hit hard by foreign competition. Workers in industries targeted by the act did benefit—that is, until they didn't. Not surprisingly, foreign nations fought back with their own slate of tariffs. In the end Smoot-Hawley only served to exacerbate the downward-spiraling economy of the Great Depression. Another example, in an inversion of sorts of the adage the road to hell is paved with good intentions, was the Immigration Act of 1924. While overtly racist in its intentions, the act had the effect of tightening up the labor market and causing upward pressure on wages, benefiting workers.[11]

Why is it that when you need Washington officialdom to be bipartisan—arresting climate change or ending gun violence—they are not? And when they are bipartisan, we could do without it.

A series of Republican and Democratic administrations from the last days of the *Fitzgerald* (Gerald Ford between 1974 and 1977) up until Donald Trump (2017–21), championed a neoliberal economic agenda that devastated the U.S. manufacturing base and by extension obliterated millions of good paying family-supporting jobs. Indeed, it was during a Democratic presidency and Democratic Congress when the innocuously titled 1978 Revenue Act was passed, establishing the 401k tax benefit in the IRS code.[12] This triggered the beginning of the end of defined benefit retirement plans (i.e., pensions) and the dawn of defined contribution benefits. In 1980 over 90 percent of private sector workers held pensions.[13] Forty years later, in 2020, less than one in five workers in the private sector held a defined-benefit retirement plan. Said Pulitzer Prize winner Gretchen Morgenson in *These are the Plunderers: How Private Equity Runs—and Wrecks—America*, "The [401 (k)] program benefitted both Wall Street and big business while exposing financial neophytes to uncertain financial futures."[14]

There was more. This phenomenon translated into additional pressures on Social Security to make up the balance of what would have been a stable income stream. Social Security was the last line of defense in retirement security, and its uncertain future triggered raw emotions. The topic became a political football that would define high profile political campaigns and attract millions of dollars of television and streaming attack ads. Rhetoric from these ads as well as speeches on the stump fueled unease in the body politic. When Social Security reform advocates prevailed in elections they were emboldened to follow through on their promises. In the wake of George W. Bush's reelection in 2004, the president's ally, Wisconsin U.S. Congressman Paul Ryan, a man who worshiped at the altar of supply side economics, drafted Social Security "reform" legislation that carved out a portion of the benefits into a 401k in which individuals could invest.[15]

Although these measures were successfully beaten back, the overall purchasing power of Social Security checks has declined precipitously. In the last twenty-three years, Social Security checks have lost 36 percent of their buying power, even though they make up a bigger portion of Americans' retirement income.[16] Today, two in five older Americans rely on Social Security exclusively for retirement income.[17]

Matters are unlikely to improve, if you listen to Americans under forty-five. Three in five adults under forty-five are either "not too confident" or "not confident at all" of the future of both Social Security and Medicare.[18] Then again, the youth vote has never had much confidence in this essential economic safety net. In 1994, a poll was conducted that showed that Generation Xers were more likely to believe in the existence of UFOs (46 percent) than that Social Security would exist when they retire (25 percent).[19]

Unlike its neighbor to the south, Canada doesn't seem to have a problem pulling together liberals and conservatives and establishing a consensus to support and grow industry and by extension create jobs. Shipbuilding is a good example.

In 2010, when conservatives held the majority in parliament, the government developed and implemented the National Shipbuilding Strategy (NSS). The policy was in response to the sector's declining fortunes. In the prior twenty-five years, the industry had slowed to a crawl. According to the government, "there had not been substantial new orders to construct vessels for many years."[20]

NSS was an industrial strategy in every sense of the term. As a thirty-year plan, it was farsighted. It had both military and nonmilitary goals. Canadians recognized that national security went well beyond a military preparedness to take in economic readiness. Both combatant and noncombatant ships were included in the package.

They put their money where their mouths were and invested substantial resources. The program was seeded with CAD$38 billion in 2011, and that number grew in subsequent years.[21]

When compared to the U.S. economy, Canada's investment in their shipbuilding industry was monumental. Canada's GDP is about one-twelfth the size of the United States, USD$1.46 trillion as opposed to USD$24 trillion. Thus, a similar program in the United States would run nearly one-half of a trillion (USD$451 billion) in nominal dollars. Currently, the value of the U.S. shipbuilding economy is USD$35.4 billion.[22]

NSS was designed to yield several social and economic benefits; the results speak for themselves. Average salaries in Canadian shipyards is 30 percent greater than the manufacturing average. It outdoes its peers, having created 2.5 times as many STEM (science, technology, engineering, and math) positions as the average of manufacturing job. Additionally, NSS provides "fair opportunities" for women and people of color.[23]

Overall, NSS has been responsible for contributing $13 billion to Canada's gross domestic product and creating or maintaining approximately twelve thousand jobs annually.[24] That level of economic development would be about a quarter of a trillion U.S. dollars in an economy the size of the United States, while creating well over one hundred thousand jobs.

The NSS is not without its critics. Opponents point to cost overruns and little bang for the buck from the billions of dollars invested in shipbuilding over the last decade. Interestingly, some of the biggest critics are from the military establishment, including *Legion Magazine*, a periodical whose mission is to "commemorate the effort, bravery and sacrifice" of both servicemen and women and veterans. The magazine recommended that Ottawa scrap the program and "consider purchasing warships from abroad."[25] Suffice it to say, it's unlikely that this phenomenon—the military establishment accusing the government of spending too much money on defense—would manifest itself in the United States. The U.S. Congress consistently gives the Defense Department more than they request in annual appropriations bills, which typically pass on wide bipartisan margins.

Meanwhile the U.S. shipbuilding economy flounders. And it flounders because U.S. policymakers are not even trying. With billions of dollars put into job training and workforce development programs along with tried-and-true apprenticeship programs at technical colleges and union halls, the United States is not without adequate infrastructure to capitalize on major investments in this sector. But this presupposes that federal policymakers *choose* to prioritize the shipbuilding industry. Elected leaders choose to outlay billions in corporate tax cuts and subsidies with no apparent policy goals and little to no evidence of job creation. Upper income tax brackets are comparatively low when put side by side with other developed countries, who typically do not have anywhere near the lucrative tax breaks that corporate America enjoys. Just take a look at Amazon's tax returns, which show an effective federal income tax rate of 6.1 percent in 2021, or ExxonMobil, whose effective rate was 2.8 percent that same year.[26] The net average tax rate for the average American worker in 2022 was 24.8 percent.[27]

Midwest industry would not be possible without a healthy lake system. A damaged Lake Superior could harm industry immeasurably. Varying water levels due to increased rainfall affects the draft of ships (i.e., how low or high they can ride). Increased storm frequency and severity means rough waters and many more chances for ships capsizing and sinking, as well as algal blooms and invasive species clogging up ballast tanks, increased maintenance costs due to lake pollution, and a less healthy environment for the sailors and crews who make the Lake Superior watershed their home. These problems will only become exacerbated as climate change worsens.

For mining and shipping companies there may be one advantage to climate change. "With global warming they're running later in the year," John O'Brien, son of Wheelsman Eugene "Red" O'Brien told me.[28] On the other hand, however, for workers that means more time away from families and fewer opportunities for fathers and sons to grow in their relationships.

To ensure Superior and the rest of the Great Lakes keep healthy, a sound land ethic must supersede all other considerations. However, if history is any indication, we have a lot of work to do. Wrote Aldo Leopold, author of *A Sand County Almanac* and one of history's most respected naturalists, "Individual thinkers since the days of Ezekiel and Isaiah have asserted that the despoliation of land is not only inexpedient but wrong. Society, however, has not yet affirmed their belief."[29]

False choices like good jobs versus a healthy environment or a working lake versus a recreational or undisturbed lake have warped the mindsets of those who make the laws, and the dichotomy has plagued public policy decisions. But there is hope.

One of the worst polluted rivers connected to the massive lake system, the lower Fox River of Wisconsin, hosts the world's largest concentration of paper mills. A heavy-polluting industry with a total lack of stewardship of the environment until the 1970s, whose communities bought into the dichotomy of good jobs versus clean environment hook, line, and sinker, the paper industry dumped an assortment of toxic and often cancer-causing chemicals into the river. Polychlorinated biphenyls (PCBs) were used in the paper-making process to make carbonless paper until 1972, when industry stopped using the carcinogenic compound. Earlier on, mills filled the river with sulphates and sulfites, byproducts from the chemical pulping process, which was responsible for widespread fish kills.[30]

It wasn't until the 1990s, when the Environmental Protection Agency browbeat the state into addressing the PCB contamination issue that PCBs were finally removed from the river or capped on its bottom. At a cost of $1.2 billion over a twenty-year period, financed entirely without public money, it was the largest cleanup of its kind in the world.[31] Today, the Fox River is well on its way to recovery. Fisheries are fully stocked, sportsmen and women can safely eat their catch, the locks system is operational for the first time in forty years, and more boaters and recreators are found in and on its waters.

On the other hand, according to a study from St. Norbert College of De Pere, Wisconsin, the community around the river suffers from unusually high cancer cases, most likely due to PCB pollution.[32] But according to data compiled by the National Cancer Institute of the CDC, the counties through which the river courses have incidence rates similar to the state (466.3 per 100,000) and the country (444.4 per 100,000). Brown County has a rate of 436.8 per one hundred thousand people; Winnebago is slightly less, at 430.0; and Outagamie is lower than the average, at 421.2.[33] Additionally, the rates themselves are falling or stable, suggesting the cleanup has improved public health outcomes.

The Great Lakes and its many tributaries are the circulatory system of its watershed and for all who reside within it. Lake Superior is its heart. Just as an animal's veins and arteries carry oxygen and important nutrients throughout its body, so too do the waters of the Great Lakes carry sustenance to nurture life and facilitate the American economy. The Ojibwe who lived in the upper Great Lakes region since time immemorial were the first to name the lake, Ojibwe Gitche Gumee or Ojibwe's Great Sea. It serves as a reminder that it is not the sea or lake of Europeans and especially not of a corporation like Reserve Mining Company or the companies that ply its waters like the former Oglebay Norton Corporation.

I made my maiden trip to Whitefish Point as I finished up the research for this book. As I looked out into Gitche Gumee from the viewing deck outside the Great Lakes Shipwreck Museum, I strained to see the point on the horizon where the *Edmund Fitzgerald* went down in 1975. My son was itching to climb down onto the beach below and look for rocks to toss into the lake, so my mind was elsewhere. George found a few flat and smooth pebbles. He wanted me to skip them across the surface of the water. George showed me how to do it.

Plunk.

Then he handed me a pebble. It must have taken hundreds if not thousands of years for the rock to make its way onto the beach, to say nothing of the pounding and scraping it endured under a two-mile-thick glacier. I ended its journey in one moment. Skip—skip—skip. It didn't come up a fourth time. I had to settle for a respectable three-skip toss.

How many times did the Fitzgerald go down and pull up before it fell to its death, I wondered. John O'Brien, who figured his father was in the pilot house when the ship went down, had turned over in his mind the last moments of his father's life countless times over the last forty-eight years. "The boat started to sink little by little. It would come up and shake itself off like a wet dog, go down, come back up, shake itself off and finally a wave hit—it was just too much. The bow took a dive and it went down, straight down, so fast."[34]

As I looked down at an unimpressed seven-year-old who was hoping for a better showing, I reminded myself that the folklore of the Fitzgerald was not really about finding cool shipwrecks and coming up with better theories as to why it sank mysteriously. Did it break apart on the surface? Was it sunk by a rogue wave? The Fitzgerald was instead about the human condition. How twenty-nine brave sailors left their families to take one more trip into the most treacherous waters and in the most harried of conditions to earn some hazard pay so they could afford an extra Christmas present for their children or maybe a bigger turkey for Thanksgiving dinner later that month. Or in the case of Bruce Hudson, a nicer engagement ring for his girlfriend who he planned to propose to at Christmas.

The late *Arthur M. Anderson* skipper Bernie Cooper downplayed the sacrifices of the crew of the Fitzgerald. "They're not martyrs to the Great Lakes. They were just a bunch of guys doing a job and they got hurt," is how Sean Ley recalled Cooper's words to me.[35] Cooper's late wife echoed similar sentiments. Deborah Champeau told me of comments Cooper's wife made to her, "Why the Fitz? There's all this commotion about the Fitz and other people have gone down in Lake Superior. Why is it the Fitz? Why is that so important?"[36]

It is understandable. The Anderson was doing the same job as the Fitzgerald. So too was the *William Clay Ford*. Both crews put their lives at risk. But Cooper's statement could not be further from the truth. And perhaps that is why the true legacy of the Fitzgerald has been lost to history for the last half century.

Part of the reason I wrote this book was to make the case to others that the crew members of the Fitzgerald were not just poor souls who took one too many trips in the worst sailing season of the year in a vessel that wasn't up to snuff. The men of the Fitzgerald were heroes. They were the reason there was enough steel to build the giant paper machines that my classmates' fathers worked on in Kaukauna or Combined Locks, Wisconsin. They were the reason John Deere of Moline, Illinois, or Case Corporation (now, Case IH) of Racine, Wisconsin, had the raw materials to build tractors and implements to pull harvests off wheat and corn fields to feed a nation. The Fitzgerald's bread and butter, taconite ore, fueled a surging manufacturing economy upon which millions of households depended for their livelihoods.

As I ended my hour-and-a-half-long call with Deborah Champeau, tears welled up in my eyes. "He'll be here, even though—but no, he's not walking me down the aisle." Powerful. Walking away from Fred Armagost's cabin outside Iron River, Wisconsin, I couldn't erase the image of an old man whose years weren't easy on him staring out into the woods like a lost soul. As Janice Armagost wavered on whether to take an interview from me, I could feel the pain in her voice and the words in her texts that betrayed a lifetime of heartache. I'd like to think their sacrifices meant more than just a successful and timeless folk-pop song. Or another facet of the greater Sault Ste Marie tourism economy. As George and I left the beach, I gazed out once more into the lake where beyond the horizon and 530 feet below the surface lies what was the *Edmund Fitzgerald*, the pride of Oglebay Norton Corporation's ship fleet.

Because the Fitz fell to its death on the Canadian side of the lake, it compelled the American families of the Fitzgerald and the Shipwreck Society to come together in common cause and prevail on the Canadian government to allow them to retrieve the ship's bell. Moreover, they recruited experts on both sides of the border to raise the bell. They used the underwater dive suit from Phil Nuytten of British Columbia; the Canadian Navy took the search party to the ship's location.

Perhaps the last chapter of the legacy of the Fitzgerald has not yet been written. It falls to all of us to remember those who served and died on it and do right by their memories. Only Janice Armagost remains among the early, vocal leaders of the families. In their places are people like Deborah Champeau and third-generation activist Ransom Cundy's grandson, Darren Rozman. As years go by, memories grow distant. This book is intended to not only recognize the crew members as heroes but also to memorialize their legacy beyond family and friends. There is no reason why this generation cannot match if not exceed expectations. Canada took a first step forward. While the National Shipbuilding Strategy is not perfect, it is the correct move. The United States can do likewise.

We must recognize what shipbuilding means: national and economic security. We got the first part right, building a Blue Water Navy that is the envy of the world. However, we cannot rest on our laurels. Countries like China are hot on our heels. Let us take inspiration from the men and women of United Steelworkers Local 8888 of Newport News Shipping, who build our nation's aircraft carriers that will be in high demand in hot spots around the world. As I write, the Fifth Fleet is ramping up presence near the Taiwanese straits as China threatens to invade the island. Additionally, Israel and Hamas are at war, and the U.S. Navy has moved considerable resources into the eastern Mediterranean Sea. Not only must we stay ahead of the competition militarily, but we also have to mobilize on the home front. That countries like South Korea can make ships faster and cheaper is no excuse not to work on our own shipbuilding capacity.

There is a clear national consensus in favor of military capital investments. Unlike other appropriation bills, the defense appropriations bill has been adopted every year by the U.S. Congress for the past sixty years.[37] The last time all appropriations bills passed on time was in 1997.[38] The bills consistently garner strong bipartisan votes. And there is plenty of money. Defense spending ($766 billion) is the fourth-largest component of the federal budget behind Social Security, health care, and income security.[39] Perhaps one economical solution is to carve out a portion of the defense budget for workforce development that could be applied to combatant and noncombatant purposes. Part of a strong shipbuilding sector is keeping workers employed and focused on their trade craft. Moving seamlessly between merchant and military ships is one solution.

Implementing a robust version of the CHIPS and Science Act of 2022 is a good model. Production costs for semiconductors are on a much lower scale, so subsidies for production of ships will have to be greater. It is a tall order. If one were to scale up Canada's NSS to match our GDP, expenditures would exceed half a trillion dollars. Overlaying the microchip legislation with Canada's NSS would ensure that industry-specific measures are included in the overall program. Think of the CHIPS and Science Act as the spirit of the policy and NSS as the brains. Reimplementing the construction differential subsidy program would be an option, too, although a return to the more robust programs of the New Deal era would be ideal.

It should also be noted that shipyards are not just on the coasts but in America's heartland, providing good-paying jobs for scores of families and building up whole communities. Fincantieri Marine of Marinette, Wisconsin, and Bay Ship of Sturgeon Bay, Wisconsin, come to mind. They have a solid labor force, and the institutional knowledge and capacity to take on additional orders, defense and nondefense projects alike. In fact, the *Arthur M. Anderson* can be found in Sturgeon Bay harbor for maintenance and has been spotted by Fitzgerald family members. On a trip to Door County in 2015, Deborah Champeau saw it. "I wanted to run over to it to hug it and kiss it for going after the Fitz in that storm," she said.[40]

Once upon a time, the shores of Lake Michigan were dotted with shipyards. Tiny Manitowoc, Wisconsin, population 34,626, just south of Green Bay, was home to its eponymous shipbuilding company, which constructed twenty-eight submarines in World War II.[41] Twenty-five went into combat and together were responsible for sinking 132 Japanese ships alone.[42] The company shuttered in 1972.

Demands for a strong shipping economy will only grow. For months after my first visit to Sault Ste Marie as an eight-year-old I pined to sail a Great Lakes ship. I also wanted to be a Green Bay Packer. We know how that panned out. My son, George, bounces from what he wants to be when he grows up like a ping-pong ball: a teacher, a Cardinal (so he can tell bishops what to do), a writer, a baseball player, or all four. Maybe shipbuilding or a job on the Great Lakes will figure in at some point. Regardless, we owe it to George's generation and future generations to make good on the legacy of the Fitzgerald by rebuilding our shipping economy and securing our economic and national security. Who knows? Maybe George will be the one to finally figure out how the *Edmund Fitzgerald* sank.

Acknowledgments

MY TRIP TO SAULT STE MARIE, AND IN PARTICULAR THE VALLEY CAMP MUSEUM THAT houses Fitzgerald artifacts including Lifeboat Number One, when I was seven years old was an awakening. It forged an indelible impression that would endure for forty years. It was the first time I truly understood the sacrifice parents make for their families. The twenty-nine Fitzgerald sailors were no different than the moms and dads I grew up with who went to the mill in Kaukauna or across the river in Kimberly and Combined Locks to work. The massive paper machines that they toiled on were not unlike the *Edmund Fitzgerald*. But there was a big difference. Unlike the millwrights and machine tenders at Thilmany or Appleton Papers, one day twenty-nine husbands and dads went to work on the Fitzgerald and never returned to their families.

This book would not be what it has become without my good friend Tim Nixon. An eleven-year veteran of the Great Lakes, he just happened to have clipped literally every news story on the Fitzgerald in the *Green Bay Press Gazette, Manitowoc Times-Herald*, *Milwaukee Journal*, and *Milwaukee Sentinel* from the day after the sinking until the wake of and fallout of the Coast Guard report. In 2017, Tim gave me the idea to challenge a receivership sale of the mill in Combined Locks, which led to a court battle and ultimately the restart of the mill and the saving of three hundred jobs. It was such a good story that I wrote a book about it. And Tim helped with that, too.

Jerald Podair is the reason I took on this project. I had assembled a few book proposals and put them in front of him for his feedback; *Wrecked* was not my first choice. Jerald responded with high praise, telling me this topic was the "perfect example" of a true historical writing. It's one thing to hear that kind of praise from a parent, quite another from one of the top writers in his field. Without his guidance and consultation through the writing process, you would not be holding this product in your hands.

My first editor Hilary Hinzmann, who plucked me from out of nowhere and took me on when I was struggling to put together my first book. I owe him a lot. Quite simply, Hilary taught me how to write.

The long list of Fitzgerald diaspora characters and family members who gave this book the color and animation that the families and readers deserve in the telling of this remarkable tale. Toby Marcovich put the first seed in my head about this project. He graciously took every call, meeting, and text I sent his way. He had much to do with the structure of Just Call Toby and many of the other chapters. Thank you, Toby!

I especially enjoyed my phone and text conversations with John O'Brien, son of Wheelsman Eugene "Red" O'Brien. John will never know how proud his father would be of the man he became.

Visiting with Fred Armagost, younger brother of Third Mate Michael Armagost on day one of my first research trip drilled home the importance of this story, that the project would be a story about the human condition. It guided me and kept me on course for the duration of the project. Thanks, Fred, for not kicking me out of your yard when I made my cold call visit. (For those who did kick me out of their yards, my sincere apologies if I offended in any way.)

Larry Tye, who blessed me with his counsel, encouragement, and advice throughout this project.

My dad, for his read-throughs and strong encouragement as well as family friends Roy and Joy. Church family, Albrents, Pautzs, Rapps and the rest of the "Packer group."

My executive assistant Anne Vane and all of Outagamie County's incredibly dedicated staff who serve our community well and make our county the best-run in the state!

Gratitude and appreciation to Bill Finan of Michigan State University Press for taking the project on and championing it every step of the way.

And finally, to Maria, my wife. And to my not-so-little-any-more Mare Mare and Georgie, who are my sunrise and sunset. (Four-legged children Barb and Charley aren't too far behind; RIP sweet Lucy.) The days are long, but the years are short, and I treasure every moment we have together. The biggest accomplishment of this book was being able to write it while making every little league game, piano practice, and play performance and that included coaching George's championship baseball team—two years in a row!

My family is a blessing, and I love them so much.

Notes

PREFACE

1. Bob Hom, interview with author, Superior, Wisconsin, June 30, 2023.
2. Carl Mueller, "Family's Ship Ties End with Tragedy," *Milwaukee Sentinel*, November 12, 2023, Tim Nixon Great Lakes Collection.
3. Bob Hom, interview with author, Superior, Wisconsin, June 30, 2023; Russell Haskell, interview with author, Sturgeon Bay, Wisconsin, July 6, 2023.
4. Fred Armagost, interview with author, Iron River, Wisconsin, June 30, 2023.
5. Lynn Williams, *One Day Stronger: A Memoir* (Ithaca, NY: ILR Press, 2011), 176.
6. Joseph MacInnis, *Fitzgerald's Storm: The Wreck of the Edmund Fitzgerald* (Holt, MI: Thunder Bay Press, 1997), 30.

CHAPTER 1. THE WRECK OF THE *EDMUND FITZGERALD*

1. Chris Foran, "What You Should Know about the Edmund Fitzgerald, the Milwaukee-Linked Ship That Sank and Inspired a Classic Gordon Lightfoot Song," *Milwaukee Journal-Sentinel*, May 3, 2023, https://www.jsonline.com/story/life/green-sheet/2023/05/03/what-to-know-about-the-edmund-fitzgerald-whose-wreck-inspired-song/70174467007/; "Edward A. Bacon Papers," Dwight D. Eisenhower Library, Accession A70–37, May 1969, https://www.eisenhowerlibrary.gov/sites/default/files/finding-aids/pdf/

bacon-edward-papers.pdf.

2. "Earth Shots: Taconite," U.S. Department of the Interior, accessed September 5, 2024, https://eros.usgs.gov/earthshots/taconite.
3. Bob Hom, interview with author, Superior, Wisconsin, June 30, 2023.
4. U.S. Department of Transportation, Coast Guard, *Marine Casualty Report: SS Edmund Fitzgerald, Sinking in Lake Superior on 10 November 1975 with Loss of Life*, Report No. USCG 16732/64216, July 26, 1977, p. 18, https://dco.uscg.mil/Portals/9/DCO%20Documents/5p/CG-5PC/INV/docs/boards/edmundfitz.pdf.
5. According to Fitzgerald cook Red Burgner, "McSorley was a rough weather-skipper. He wouldn't haul up for nothing. We're going through, he said. We don't make no money sitting at anchor." Ric Mixter (dir.), *Expedition '94: To the Fitzgerald* (documentary), Flint, MI, 1995.
6. U.S. Department of Commerce, National Oceanic and Atmospheric Administration, "Storm Warning: Advancements in Marine Forecasting since the Edmund Fitzgerald," National Weather Service, accessed September 4, 2024, https://www.weather.gov/mqt/fitz_fitza#.
7. Michael Schumacher, *Mighty Fitz: The Sinking of the Edmund Fitzgerald* (New York: Bloomsbury, 2005), 27–28.
8. Schumacher, *Mighty Fitz*, 7.
9. Kay McSorley, phone interview with author, August 5, 2023.
10. Toby Marcovich, interview with author, Superior, Wisconsin, June 30, 2023.
11. Schumacher, *Mighty Fitz*, 26.
12. "Interview with Cheryl Rozman," SS Edmund Fitzgerald Online, February 20, 2005, https://ssedmundfitzgerald.org/cheryl-rozman-interview.
13. Michael Schumacher, *The Trial of the Edmund Fitzgerald* (Minneapolis: University of Minnesota Press, 2019).
14. Schumacher, *Trial*, 4.
15. Schumacher, *Trial*, 10.
16. Tim Nixon, interview with author, Green Bay, Wisconsin, July 21, 2023.
17. Jeff Brown, "Nearly 50 Years of Occupational Safety and Health Data," *Beyond the Numbers* 9, no. 9 (July 2020), https://www.bls.gov/opub/btn/volume-9/nearly-50-years-of-occupational-safety-and-health-data.htm; David Rosner and Gerald Markowitz, "A Short History of Occupational Safety and Health in the United States," *American Journal of Public Health* 110, no. 5 (May 2020): 622–28; https://doi.org/10.2105/AJPH.2020.305581.
18. Tim Nixon, interview with author, Green Bay, Wisconsin, July 21, 2023.
19. Schumacher, *Trial*, 6.

20. She sailed seventeen seasons and made 750 round trips in total (Melanie Chida, "The Wreck of the SS Edmund Fitzgerald: The Fitzgerald's Last Journey across Lake Superior," ArcGIS StoryMaps, December 7, 2021, https://storymaps.arcgis.com/stories/b8d4d97131b34da6af8101bb9caecd6f).
21. Judith Stein, *Pivotal Decade* (New Haven: Yale University Press) 2010; Judith Stein, *Running Steel, Running America* (Chapel Hill: University of North Carolina Press) 1998, 214; Regarding industrial strategy proposals of the late 1970s, see Seymour Zucker, *The Reindustrialization of America* (New York: McGraw-Hill) 1982.
22. Schumacher, *Trial*, 18.
23. Ibid.
24. Schumacher, *Trial*, 18–19.
25. U.S. Department of Transportation, *Marine Casualty Report*, 22.
26. U.S. Department of Transportation, *Marine Casualty Report*, 26.
27. Jim Woodward was unsure if it was the first or second mate that was denied permission to call the Fitzgerald. According to the Coast Guard report, the second mate had assumed watch at noon and would hold watch until 4 p.m. (Jim Woodward, phone interview with author, August 5, 2023).
28. Sean Ley, "The Fateful Journey," Great Lakes Shipwreck Museum, accessed September 4, 2024, https://shipwreckmuseum.com/the-fateful-journey/.
29. Mixter would go on to argue that there was no such thing as the Six Fathom Shoal. The "1" in 16 (fathoms) on a chart was obscured and the area in question was actually 16 fathoms, deep enough for the Fitzgerald to clear even in rough waters. Ric Mixter, *Tattletale Sounds: The Edmund Fitzgerald Investigations* (Wixom, MI: Airworthy Productions, 2022), 125.
30. Tanda Gmiter, "This Timeline Chronicles the Last Voyage of the Edmund Fitzgerald," MLive, November 9, 2018. https://www.boreal.org/2018/11/09/183767/this-timeline-chronicles-the-last-voyage-of-the-edmund-fitzgerald.
31. Ric Mixter, "Tattletale Sounds: The Edmund Fitzgerald Investigations," Bay County Historical Society, YouTube, posted January 18, 2023, https://youtu.be/OW-tig-2P-I.
32. Jeff Thomas, "We are Holding Our Own: Edmund Fitzgerald," Fresh Water Bible Church, YouTube, posted March 31, 2023, https://www.youtube.com/live/rZoJmAX09Hs.
33. "Remembering the Edmund Fitzgerald," *The Daily Mining Gazette*, November 11, 2016, https://www.mininggazette.com/news/local-news/2016/11/remembering-the-edmund-fitzgerald/.
34. Schumacher, *Trial*, 32.
35. Jim Hettinger, "Remembering Capt. J.B. Cooper," *Battle Creek Enquirer*, October 8, 2016.

https://www.battlecreekenquirer.com/story/opinion/columnists/2016/10/08/jim-hettinger-remembering-capt-jb-cooper/91752270/.

36. Gmiter, "This Timeline."
37. Schumacher, *Trial*, 32.
38. Schumacher, *Trial*, 33.
39. Ibid.

CHAPTER 2. THE WRECK OF THE AMERICAN ECONOMY

1. John Cassidy, "Piketty's Inequality Story in Six Charts," *New Yorker*, March 26, 2014, https://www.newyorker.com/news/john-cassidy/pikettys-inequality-story-in-six-charts.
2. Christopher Ingraham, "Union membership remained steady in 2017. The trend may not hold," *Washington Post*, January 19, 2018, https://www.washingtonpost.com/news/wonk/wp/2018/01/19/union-membership-remained-steady-in-2017-the-trend-may-not-hold/; Visualizing Economics, "Average income in the United States," 2008; Justin Wolfers, "The Unemployment Rate is Probably Around 13 Percent," *New York Times*, April 3, 2020; Jack Metzger, *Striking Steel* (Philadelphia: Temple University Press, 2000), 30.
3. Franklin D. Roosevelt Library and Museum, "Great Depression," https://www.fdrlibrary.org/great-depression-facts; Macrotrends, "Dow Jones—100 Year Historical Chart," https://www.macrotrends.net/1319/dow-jones-100-year-historical-chart.
4. Long Legacy Project, "Share Our Wealth," Huey Long, 2024, https://www.hueylong.com/programs/share-our-wealth.php.
5. Ira Katznelson, *Fear Itself* (New York: Liveright, 2013), 35.
6. Nelson Lichtenstein, *State of the Union: A Century of American Labor* (Princeton, NJ: Princeton University Press, 2003), 64.
7. Wendell Cox, "500 Years of GDP: A Tale of Two Countries," *New Geography*, September 21, 2015, https://www.newgeography.com/content/005050-500-years-gdp-a-tale-two-countries.
8. Alan Brinkley, *The End of Reform: New Deal Liberalism in Recession and War* (New York: Knopf, 1996), 38.
9. James A. Gross, *Broken Promises: The Subversion of US Labor Relations* (Philadelphia: Temple University Press, 1995), 1.
10. H. W. Brands, *Traitor to His Class: The Privileged Life and Radical Presidency of Franklin Delano Roosevelt* (New York: Knopf, 2000), 436.
11. The National World War II Museum, "Research Starters: US Military by the Numbers," National World War II Museum, accessed September 4, 2024, https://www.

nationalww2museum.org/students-teachers/student-resources/research-starters/research-starters-us-military-numbers.

12. The American Presidency Project, "Franklin D. Roosevelt: Campaign Address on Progressive Government at the Commonwealth Club in San Franciso, California," American Presidency Project, September 23, 1932, https://www.presidency.ucsb.edu/documents/campaign-address-progressive-government-the-commonwealth-club-san-francisco-california.
13. The American Presidency Project, "Franklin D. Roosevelt: State of the Union Message to Congress," American Presidency Project, January 11, 1944, https://www.presidency.ucsb.edu/documents/state-the-union-message-congress.
14. National Park Service, "Lincoln's Second Inaugural Address," National Park Service, April 18, 2020, https://www.nps.gov/linc/learn/historyculture/lincoln-second-inaugural.htm.
15. Harry S. Truman Library and Museum, "Steel Strike of 1952," Truman Library, accessed September 4, 2024, https://www.trumanlibrary.gov/education/presidential-inquiries/steel-strike-1952.
16. Youngstown Sheet & Tube Co. v. Sawyer, 343 U.S. 579 (1952).
17. Frederick R. Barkley, "Full Jobs Measure Assailed by Taft," *New York Times*, August 30, 1945.
18. Jon Shelton, phone interview with author, May 19, 2024.
19. Federal Reserve History "Employment Act of 1946" February 20, 1946.
20. Patrick Christy, "America's Proud History of Post-War Aid," *US News and World Report*, June 6, 2014.
21. Europe Direct. "1950–2020: 70 years of the Schuman Declaration and of the great strategy of "small steps." https://www.strasbourg-europe.eu/the-schuman-declaration
22. Rachel Premack, "How America Quietly Lost 2,700 Ships," Business Insider, June 18, 2021, https://www.businessinsider.com/how-america-quietly-lost-2700-ships-maritime-dominance-2021-6.
23. Nelson Lichtenstein, *Walter Reuther: The Most Dangerous Man in Detroit* (Urbana: University of Illinois Press, 1997), 280.
24. See Elizabeth Tandy Shermer, *Sunbelt Capitalism: Phoenix and the Transformation of American Politics* (Philadelphia: University of Pennsylvania Press, 2013).
25. Statista, "Aggregated CEO-to-Worker Compensation Ratio for the 350 Largest Publicly Owned Companies in the United States from 1965 to 2022," accessed September 4, 2024, https://www.statista.com/statistics/261463/ceo-to-worker-compensation-ratio-of-top-firms-in-the-us/; Matthew Smith, Owen Zidar, and Eric Zwick, "Top Wealth in America: New Estimates under Heterogeneous Returns," Princeton Economics, October 2021,

https://economics.princeton.edu/working-papers/top-wealth-in-america-new-estimates-under-heterogenous-returns/.

26. Okazaki Tetsuji, "Lessons from the Japanese Miracle: Building the Foundations for a New Growth Paradigm," Nippon, February 9, 2015, https://www.nippon.com/en/in-depth/a04003/.
27. For a comprehensive and fascinating rendering of this geopolitical competition, read Deborah Cadbury, *Space Race: The Epic Battle between America and Soviet Union for Dominion of Space* (New York: Harper, 2006).
28. NASA Technology Transfer Program, "NASA Spinoff," NASA, accessed September 4, 2024, https://spinoff.nasa.gov/.
29. Arta Moeini and Alexandra Slobodov, "Rethinking Nuclear Arms Control for a Multipolar World," Institute for Peace and Diplomacy, October 23, 2020, https://peacediplomacy.org/2020/10/23/rethinking-nuclear-arms-control-for-a-multipolar-world/.
30. Lawrence Mishel and Jessica Schieder, "As Union Membership Has Fallen, the Top 10 Percent Have Been Getting a Larger Share of Income," Economic Policy Institute, May 24, 2016, https://www.epi.org/publication/as-union-membership-has-fallen-the-top-10-percent-have-been-getting-a-larger-share-of-income/.
31. Gabriel Winant, *The Next Shift* (Chicago: University of Chicago Press, 2021), 11.
32. Karly Schledwitz, "Historic Political Blackmail Led to Senate Suicide and Delay of Gay Rights," NBC News, June 10, 2013, https://www.nbcnews.com/id/wbna52158960.
33. "Work Stoppages (Strikes and Lockouts)," Digital History Reader, accessed September 4, 2024, https://www.dhr.history.vt.edu/modules/us/mod06_1936/images/WorkStoppages.jpg.
34. Colin Gordon and Ross Eisenbrey, "As Unions Decline, Inequality Rises," Economic Policy Institute, June 6, 2012, https://www.epi.org/publication/unions-decline-inequality-rises/.
35. Johnson once commissioned a study by the Library of Congress to prove Woodrow Wilson was not from the South. (Wilson was born in Virginia and grew up in Georgia.) Nicholas Lehman, *The Promised Land: The Great Black Migration and How it Changed America* (New York: Knopf, 1991), 137; Lyle C. Wilson, "Washington Window: Johnson First Southern President since Andrew Johnson," November 23, 1963, https://www.upi.com/Archives/1963/11/23/Washington-Window-Johnson-first-southern-president-since-Andrew-Johnson/9415458482043/.
36. Martin Luther King Jr., "*All Labor Has Dignity*," ed. Michael K. Honey (Boston: Beacon Press, 2011), 35.
37. Nisha Bodway, "From the Dean: Reflections on the Legacy of Hubert Humphrey's 1948 Civil Rights Speech," Hubert H. Humphrey School of Public Affairs. University of

Minnesota. July 14, 2023.

38. Larry Tye, "Bobby Kennedy: The Making of a Liberal Icon," John F. Kennedy Library Foundation, YouTube, October 5, 2016, https://www.youtube.com/live/5IBHP6_g4Bk.
39. Brooks Jackson, "Blacks and the Democratic Party," FactCheck.org, April 18, 2008, https://www.factcheck.org/2008/04/blacks-and-the-democratic-party/.
40. Juliette Cubanski, Jeannie Fulgesten Biniek, and Tricia Neuman, "FAQs on Health Spending, the Federal Budget, and Budget Enforcement Tools," KFF, March 20, 2023, https://www.kff.org/medicare/issue-brief/faqs-on-health-spending-the-federal-budget-and-budget-enforcement-tools/.
41. NASA, "Vital Signs: Carbon Dioxide Measurements," NASA, updated July 2024, https://climate.nasa.gov/vital-signs/carbon-dioxide/.

CHAPTER 3. THE LAKE THEY CALL GITCHE GUMEE

1. Gaylord Nelson, Earth Day Speech, The Woman's Club of Minneapolis. April 21, 1994, https://archive.mpr.org/stories/1994/04/21/gaylord-nelson-environmental-speech-followed-by-dean-abrahamson-discussing-earth-day
2. Eva Ryan, "Ojibwe Stories of Spirit Island," Sea Grant (blog), University of Wisconsin, January 31, 2022, https://www.seagrant.wisc.edu/blog/ojibwe-stories-of-spirit-island/.
3. Marie Zhuikov, "Wisconsin Researchers Study 'Rogue Waves' Like Ones Thought to Have Sunk the Fitzgerald," Sea Grant, University of Wisconsin, November 9, 2015, https://www.seagrant.wisc.edu/news/wisconsin-researchers-study-rogue-waves-like-ones-thought-to-have-sunk-the-fitzgerald/.
4. National Parks of Lake Superior Foundation, "Lake Superior Shipwrecks: Explore 8 Famous Wrecks," NPLSF, accessed September 4, 2024, https://www.nplsf.org/post/lake-superior-shipwrecks.
5. Nancy Langston, *Sustaining Lake Superior: An Extraordinary Lake in a Changing World* (New Haven, CT: Yale University Press. 2017), 2.
6. "The Great Lakes Seaway Navigation System." Amerian Great Lakes Port Assocciation. https://www.greatlakesports.org/industry-overview/the-great-lakes-seaway-navigation-system/
7. Louis G. Williams, "Should Some Beneficial Uses of Public Waterways Be Illegitimate?" *BioScience* 18, no. 1 (1968): 36–37, quoted in Langston, *Sustaining Lake Superior*.
8. Sheri McWhirter, "Part of Tribal Identity, Great Lakes Wild Rice Threatened by Climate Change," MLive, October 6, 2022, https://www.mlive.com/public-interest/2022/10/part-of-tribal-identity-great-lakes-wild-rice-threatened-by-climate-change.html.

9. Jarrod T. Hardke, "Water-Use Efficiency Options," Rice Farming, April 1, 2013, https://www.ricefarming.com/departments/specialists-speaking/water-use-efficiency-options/.
10. Chantal Norrgard, *Seasons of Change* (Chapel Hill: University of North Carolina Press, 2014), 23, 25.
11. Grace Lee Nute, *Lake Superior* (Minneapolis: University of Minnesota Press, 1944), x.
12. Nute, *Lake Superior*, 45. Because of dialectic differences, "Gitche Gumee" can be spelled more than one way.
13. Nute, *Lake Superior*, 69.
14. "Ely Parker: Seven Generations—the Role of Chief," PBS, https://www.pbs.org/warrior/content/timeline/opendoor/roleOfChief.html.
15. Jennifer Sass, "Scientific Evidence to Support 'Seven Generations' Future Thinking" NRDC (blog), November 28, 2013, https://www.nrdc.org/bio/jennifer-sass/scientific-evidence-support-seven-generations-future-thinking-our-toxic-chemical. Examples of studies include Carlos Guerrero-Bosagna, Marina Savenkova, Muksitul Haque, Eric Nilsson, and Michael K. Skinner, "Environmentally Induced Epigenetic Transgenerational Inheritance of Altered Sertoli Cell Transcriptome and Epigenome: Molecular Etiology of Male Infertility," *PLOS ONE 8*, no. 3 (2013): e59922, https://doi.org/10.1371/journal.pone.0059922; Mohan Manikkam, Carlos Guerrero-Bosagna, Rebecca Tracey, Muksitul Haque, and Michael K. Skinner, "Transgenerational Actions of Environmental Compounds on Reproductive Disease and Identification of Epigenetic Biomarkers of Ancestral Exposures," *PLOS ONE 7*, no. 2 (2012): e31901, https://doi.org/10.1371/journal.pone.0031901.
16. U.S. Department of State, Office of the Historian, "The Truman Doctrine and the Marshall Plan," Office of the Historian, accessed November 3, 2023, https://history.state.gov/departmenthistory/short-history/truman.
17. Langston, *Sustaining Lake Superior*, 79.
18. Langston, *Sustaining Lake Superior*, 115.
19. Evidently this was such a compelling statistic that the author cited the number no fewer than three times in the text (Langston, *Sustaining Lake Superior*, 81, 85, 88).
20. Langston, *Sustaining Lake Superior*, 85.
21. Langston, *Sustaining Lake Superior*, 87.
22. Langston, *Sustaining Lake Superior*, 102.
23. Langston, *Sustaining Lake Superior*, 107.
24. Ric Mixter, *Tattletale Sounds: The Edmund Fitzgerald Investigations* (Wixom, MI: Airworthy Productions, 2022), 14.

25. Peter Schilling Jr., "Hard-Fought United States vs. Reserve Mining Changed Environmentalism," *MinnPost*, February 15, 2013, https://www.minnpost.com/minnesota-history/2013/02/hard-fought-united-states-vs-reserve-mining-changed-environmentalism/.
26. Langston, *Sustaining Lake Superior*, 136.
27. Langston, *Sustaining Lake Superior*, 98.
28. For a comprehensive accounting of the effects of climate change on Lake Superior, see Langston, *Sustaining Lake Superior*, chap. 9. This section summarizes the fallout.

CHAPTER 4. ROCK BOTTOM

1. Thomas G. Lubenow, "Little Hope Seen for Ship's Crew," *Milwaukee Journal*, November 12, 1975.
2. Ric Mixter, *Tattletale Sounds: The Edmund Fitzgerald Investigations* (Wixom, MI: Airworthy Productions, 2022), 98.
3. "Lake Freighter Sinks in Storm," *Green Bay Press Gazette*, November 11, 1975.
4. Tim Nixon, interview with author, Green Bay, Wisconsin, July 21, 2023.
5. A partial transmission exchange between the Coast Guard and *Anderson* can be found https://shipwreckmuseum.com/the-fateful-journey/.
6. Ibid.
7. Tim Nixon knew the brother and sailed with him in the 1979 season. At the time, the brothers were still not on speaking terms (Tim Nixon, interview with author, Green Bay, Wisconsin, July 21, 2023).
8. Later, other ships would respond. The *Arthur A. Anderson* and *William Clay Ford* were the only ones to respond in the affirmative in the first round. Fred Stonehouse, *The Wreck of the Edmund Fitzgerald* (Gwinn, MI: Avery Color Studios, 1999), 41–42.
9. "Fitzgerald Was in Good Repair," *Green Bay Press Gazette*, November 13, 1975.
10. Tim Nixon, interview with author, Green Bay, Wisconsin, July 21, 2023.
11. Michael Schumacher, *The Mighty Fitz: The Sinking of the Edmund Fitzgerald* (New York: Bloomsbury, 2005), 96.
12. "No Trace of 29 in Ship Sinking," *Milwaukee Sentinel*, November 12, 1975.
13. "Wreckage Might Be Fitzgerald," *Herald-Times Reporter*, November 14, 1975.
14. "Detect Wreckage in Lake Superior," *Herald-Times Reporter*, November 17, 1975.
15. "Fitzgerald Resting Spot Confirmed," *Milwaukee Sentinel*, November 27, 1975.
16. "Fitzgerald Was in Good Repair," *Green Bay Press Gazette*, November 13, 1975.

17. "SS Edmund Fitzgerald Sinking in Lake Superior, Marine Accident Report, National Transportation Safety Board," Michigan Sea Grant, May 4, 1978, p. 11, https://www.michiganseagrant.org/downloads/lessons/datasets/earthscience/Edmund-Fitzgerald-accident-report-NTSBR.pdf.
18. Fred Stonehouse, *The Wreck of the Edmund Fitzgerald* (Gwinn, MI: Avery Color Studios, 2006), 121.
19. Schumacher, *Mighty Fitz*, 120–21.
20. "29 Feared Lost as Ship Sinks in Lake Superior," *Wisconsin State Journal*, November 12, 1975.
21. "No Trace of 29 in Ship Sinking," *Milwaukee Sentinel*, November 12, 1975.
22. Lubenow, "Little Hope."
23. "Lake May Clutch Victims Forever," *Green Bay Press Gazette*, November 12, 1975.
24. There was debate whether or not air compartments such as sleeping quarters or the mess hall would collapse under immense water pressure at five hundred feet. However, even if they were trapped within the walls of their rooms or elsewhere, they would no doubt succumb to the frigid temperatures within hours if not minutes.
25. "Inquiry Planned into Ship Sinking," *Milwaukee Sentinel*, November 14, 1975.
26. "Sonar Hints Fitzgerald Split in Two," *Milwaukee Sentinel*, November 19, 1975.
27. "Union Protests as Probe of Sinking Opens," *Green Bay Press-Gazette*, November 18, 1975.
28. Mixter, *Tattletale Sounds*, 120–23.
29. "Lake Bottom Wreck Likely Is Fitzgerald," *Milwaukee Sentinel*, November 17, 1975.
30. "Sonar Hints Fitzgerald Split in Two," *Milwaukee Sentinel*, November 19, 1975.
31. "4-Man Board Probes Sinking of Ore Carrier," *Herald-Times Reporter*, November 18, 1975.
32. "One Court Sought in Fitzgerald Case," *Milwaukee Sentinel*, December 3, 1975.
33. "Pilot of Boat near Fitzgerald Will Give Testimony Wednesday," *Herald Times Reporter*, November 25, 1975.
34. "Another Tragedy on the Lakes," *Steel Labor*, December 1975.
35. Ibid.
36. Richard Bradee, "US Safety Board Rejects Fitzgerald Sinking Report," *Milwaukee Sentinel*, March 24, 1978.
37. Mixter, *Tattletale Sounds*, 166.
38. Mixter, *Tattletale Sounds*, 162.
39. U.S. Department of Transportation, Coast Guard, *Marine Casualty Report: SS Edmund Fitzgerald, Sinking in Lake Superior on 10 November 1975 with Loss of Life*, Report No. USCG 16732/64216, July 26, 1977, p. 104, https://dco.uscg.mil/Portals/9/DCO%20

Documents/5p/CG-5PC/INV/docs/boards/edmundfitz.pdf.

40. "Inquiry Pins Fate of 'Fitz' to Faulty Hatches," *Detroit News*, August 2, 1977.

CHAPTER 5. JUST CALL TOBY

1. Deborah Champeau, phone interview with author, August 11, 2023.
2. Sources: Toby Marcovich, interview with author, Superior, Wisconsin, June 30, 2023; Dick Hunegs, interview with author, St. Louis Park, Minnesota, August 19, 2023; Dick Theno, interview with author, Maplewood, Minnesota, August 20, 2023; Steve Eckman, interview with author, Independence, Minnesota, September 9, 2023; Peter Riley, phone interview with author, July 31, 2023; Peter Riley, "The Sinking of the Edmund Fitzgerald: The Untold Story" (PowerPoint presentation); "Burgner Deposition," [*Minneapolis*] *Star Tribune*, November 10, 1997.
3. Toby Marcovich, interview with author, Superior, Wisconsin, June 30, 2023; U.S. Bureau of Labor Statistics, "CPI Inflation Calculator," https://data.bls.gov/cgi-bin/cpicalc.pl?cost1=1.00&year1=193402&year2=202306.
4. "Impact of Tort Reform on Personal Injury Lawsuits," Justia, updated October 2023, https://www.justia.com/injury/negligence-theory/tort-reform/.
5. Thomas F. Burke, *Lawyers, Lawsuits, and Legal Rights: The Battle Over Litigation in American Society* (Berkeley: University of California Press, 2002), 52.
6. It is unclear whether Oglebay Norton Corporation or Northwestern Mutual had responded to legal actions. Some of the attorneys, legal investigators, and family members that I interviewed had difficulty discerning one from the other or gave conflicting accounts.
7. Tim Nixon, interview with author, Green Bay, Wisconsin, July 21, 2023; Frank Maraist and Thomas Galligan, "Chapter X: General Average," in *Admiralty in a Nutshell*, 6th ed. (Saint Paul, MN: West Academic Publishing, 2010).
8. "Repeat of Tragedy," *Steel Labor*, December 1975.
9. Tim Nixon, interview with author, Green Bay, Wisconsin, July 21, 2023; Toby Marcovich, interview with author, Superior, Wisconsin, June 30, 2023; Dick Hunegs, interview with author, St. Louis Park, Minnesota, August 19, 2023.
10. James Haskell, interview with author, Sturgeon Bay, Wisconsin, July 6, 2023.
11. Robert O. Self, *All in the Family: The Realignment of American Democracy Since the 1960s* (New York: Farrar, Straus and Giroux, 2013), 7.
12. Toby Marcovich, interview with author, Superior, Wisconsin, June 30, 2023.
13. Steve Eckman, interview with author, Independence, Minnesota, September 9, 2023.

14. Mileage estimate is based on flight hours logged per Toby Marcovich and the average speed of a Cessna airplane (160knots/140 mph).
15. U.S. Department of Transportation, Coast Guard, *Marine Casualty Report: SS Edmund Fitzgerald, Sinking in Lake Superior on 10 November 1975 with Loss of Life*, Report No. USCG 16732/64216, United States Coast Guard, July 26, 1977, https://www.dco.uscg.mil/Portals/9/DCO%20Documents/5p/CG-5PC/INV/docs/boards/edmundfitz.pdf.
16. There was one exception. It was tradition, according to attorney Steve Eckman, for all crew members and officers to mud-out annually; that is, to scoop out the sludge and sediment that had accumulated in the hull over the season. It was akin to spring cleaning in a home when parents and kids alike pitch in. Incidentally, it was this tradition that placed Red Burgner, a cook, in the bottom hold and discovered the tack-welded steel on the keel.
17. Brady Slater, "Forty years after the sinking of the Fitzgerald, untold stories emerge," *Duluth News Tribune*, November 7, 2015.
18. For more information on the case, see *Cederberg v. Robison*, 100 Cal. 93, 34 P. 625 (Cal. 1893), https://casetext.com/case/cederberg-v-robison.
19. Dick Theno, interview with author, Maplewood, Minnesota, August 20, 2023.
20. Dick Hunegs, interview with author, St. Louis Park, Minnesota, August 19, 2023.
21. Reproductions of the Burgner deposition were acquired via two sources: *Deposition of George H. Burgner Regarding the Edmund Fitzgerald* (Bowling Green, OH: Bowling Green State University Library, Great Lakes Collection, 2015), https://researchworks.oclc.org/archivegrid/collection/data/907379010 and Dick Theno, "The 26th Infantry Division During World War II," Holocaust Encyclopedia, https://encyclopedia.ushmm.org/content/en/article/the-26th-infantry-division.
22. "Burgner Deposition."
23. Ric Mixter, *Tattletale Sounds: The Edmund Fitzgerald Investigations* (Wixom, MI: Airworthy Productions, 2022), 156.
24. "Burgner Deposition."
25. Steve Eckman, interview with author, Independence, Minnesota, September 9, 2023.
26. "Gordon Lightfoot: Chart History," Billboard Hot 100, accessed November 4, 2023, https://www.billboard.com/artist/gordon-lightfoot/.
27. Dick Hunegs, interview with author, St. Louis Park, Minnesota, August 19, 2023.
28. "Burgner Deposition."
29. Attorneys and investigators of the DeParque firm either did not recall the amount of the settlements or refused to disclose on the record. However, one confirmed that the Kalmon settlement was larger than Wilhelm's because the latter had more dependents

and greater earning potential. The Kalmon figure came from "Burgner Deposition," [*Minneapolis*] *Star Tribune*, November 10, 1997.

30. Toby Marcovich, interview with author, Superior, Wisconsin, June 30, 2023.
31. Deborah Champeau, phone interview with author, August 11, 2023.

CHAPTER 6. A BONE TO BE CHEWED

1. Greg Quill, "Gordon Lightfoot Changes Edmund Fitzgerald Lyrics," *Toronto Star*, March 25, 2010, https://www.thestar.com/entertainment/music/gordon-lightfoot-changes-edmund-fitzgerald-lyrics/article_dc683325-9aea-574d-82f6-992a5b5e6e21.html.
2. U.S. Department of Transportation, Coast Guard, *Marine Casualty Report: SS Edmund Fitzgerald, Sinking in Lake Superior on 10 November 1975 with Loss of Life*, Report No. USCG 16732/64216, United States Coast Guard, July 26, 1977, https://dco.uscg.mil/Portals/9/DCO%20Documents/5p/CG-5PC/INV/docs/boards/edmundfitz.pdf.
3. "Launch of the Great Lakes Carrier SS Edmund Fitzgerald June 7, 1958," Internet Archive, August 1, 2017, https://archive.org/details/56474EFitzgeraldLaunchMos.
4. Skip Meier and Wayne Garrett, *Great Lakes Engineering Works: The Shipyard and Its Vessels* (Detroit, MI: The Marine Historical Society of Detroit, 2008), https://www.mhsd.org/glew/1shipyard.pdf.
5. "Democrats Sweep 1958 Elections; Will Have 64 Senators, 283 Representatives, 35 Governors," *CQ Almanac 1958*, 14th ed., 11-713-11-716 (Washington, DC: Congressional Quarterly, 1959), http://library.cqpress.com/cqalmanac/cqal58-1340275.
6. "The Patrick Vincent McNamara Collection," Walter P. Reuther Library, Wayne State University, Detroit, Michigan, https://reuther.wayne.edu/files/UP000189.pdf.
7. Deborah Champeau, phone interview with author, August 11, 2023.
8. Reproductions of the Burgner deposition were acquired via two sources: *Deposition of George H. Burgner Regarding the Edmund Fitzgerald* (Bowling Green, OH: Bowling Green State University Library, Great Lakes Collection, 2015), https://researchworks.oclc.org/archivegrid/collection/data/907379010 and Dick Theno, interview with author, Maplewood, Minnesota, August 20, 2023.
9. Dustin Delsman, text with author, November 14, 2024.
10. Northwestern Mutual outside counsel Thomas Murphy objected to this portion of the testimony and other portions of the testimony when Burgner refers to comments made to him by others. It should also be noted that when attorney Steve Eckman, counsel to two Fitzgerald families, concluded his questioning and turned the witness over to Murphy, both Murphy and Oglebay Norton Corporation personnel manager Robert

Kratzert declined to ask Burgner *any* questions despite two and a half hours of testimony that included multiple objections on the part of Murphy.

11. Mike TenEyck credentials: Glenn Rifkin, "A Great Lakes Shipping Disaster," Briefings Magazine, Korn Ferry, https://www.kornferry.com/insights/briefings-magazine/issue-59/a-great-lakes-shipping-disaster; Mike TenEyck, phone interview with author, October 3, 2023. TenEyck declined to share his white paper with me.
12. Mike TenEyck, phone interview with author, October 3, 2023.
13. Ibid.
14. Association of Flight Attendants, "Flight Attendant Union Statement on the Boeing 737 MAX Return to Service" (news release), AFA, November 18, 2020, https://www.afacwa.org/statement_boeing_737max_return_service.
15. Michael Schumacher, *The Trial of the Edmund Fitzgerald* (Minneapolis: University of Minnesota Press, 2019), 23.
16. Mike TenEyck, phone interview with author, October 3, 2023.
17. Bob Hom, interview with author, Superior, Wisconsin, June 30, 2023.
18. The Fitzgerald was not the only ship that Oglebay Norton Corporation maintained on the cheap. The SS *Sylvania*, which Oglebay Norton Corporation owned from 1971 until it was scrapped in 1980, suffered six accidents including an incident in 1973 when a propeller literally fell off the vessel. To put this into perspective, the ship had about as many reported incidents (7) that required repairs over its previous sixty-five-year history (1905–70) (https://greatlakes.bgsu.edu/item/440048).
19. "Burgner deposition." In other accounts, Burgner refers to Red Mead using a coal shovel and not a crowbar to show him the broken weld.
20. "Burgner deposition."
21. OSHA Directive Number: CPL-02-01-020, https://osha.gov/enforcement/directives/cpl-02-01-020.
22. "Burgner deposition."
23. Ibid.
24. In the deposition and *Expedition '94: To the Fitzgerald!* documentary, Burgner refers to Thompson as the president of Oglebay Norton Corporation even though the record shows he was a senior vice president (1972–73) and executive vice president of operations (1973–81) at the time. Thompson would rise to the position of president and CEO in 1982 ("Renold Durant Thompson," Prabook, https://prabook.com/web/renold_durant.thompson/373941).
25. In the transcript, Burgner refers to Bill as a watchman. I deduced it to be William Spengler because he was the only "William" or "Bill" on the crew, and he was a watchman.

26. It would seem that in addition to having a longshoreman or seafaring license, bearing the nickname "Red" would also qualify you to work on the Great Lakes. Although he was not present, John O'Brien, son of Red O'Brien, thought it unlikely that his dad would have called attention to a structural problem. The two spent a lot of time together in the off season, and John "never heard him complain about anything" (John O'Brien, phone interview with author, August 28, 2023).
27. Ric Mixter (dir.), *Expedition '94: To the Fitzgerald*. Flint, Michigan, 1995. Burgner was consistent in this recollection. He had offered a similar account in his deposition seventeen years earlier in December of 1977.
28. Bethany McLean and Joe Nocera, *All the Devils Are Here: The Hidden History of the Financial Crisis* (New York: Portfolio, 2011).
29. "Moody's $864 Million Penalty for Ratings in Run-Up to 2008 Financial Crisis," *UK Guardian*, January 14, 2017.
30. Renae Merle, "A Guide to the Financial Crisis 10 Years Later," *Washington Post*, September 10, 2018, https://www.washingtonpost.com/business/economy/a-guide-to-the-financial-crisis--10-years-later/2018/09/10/114b76ba-af10-11e8-a20b-5f4f84429666_story.html.
31. Kim Phillips-Fein, *Fear City: New York's Fiscal Crisis and the Rise of Austerity Politics* (New York: Metropolitan Books, 2017), 43, 89.
32. Mixter, *Expedition '94*.
33. Ibid.
34. Tim Nixon, interview with author, Green Bay, Wisconsin, July 21, 2023.
35. Richard Orgel to Robert W. Graham, April 3, 2015, In *Deposition of George H. Burnger Regarding the Edmund Fitzgerald* (Bowling Green, OH: Bowling Green State University Library, Great Lakes Collection, 2015).
36. The index of the deposition at the Bowling Green University archive was misspelled. See "Burgner deposition."

CHAPTER 7. A RED SEA

1. "US Top 40 Singles Week Ending 16th October 1976," Top 40 Weekly, https://top40weekly.com/2013/10/16/us-top-40-singles-week-ending-16th-october-1976/; Michael Levenson, "Bill Fries, Singer Known for 1970s Trucking Ballad 'Convoy,' Dies at 93," *New York Times*, April 3, 2022.
2. "JFK Assassination Records, Summary of Findings," National Archives, August 15, 2016, https://www.archives.gov/research/jfk/select-committee-report/summary.html.
3. Meghan Keneally, "Why the Apollo 11 Moon Landing Conspiracy Theories Have Endured

Despite Being Debunked Numerous Times," ABC News, July 18, 2019, https://abcnews.go.com/Technology/apollo-11-moon-landing-conspiracy-theories-endured-debunked/story?id=64339363; "Did Astronauts Really Land on the Moon?" NASA Facts, February 14, 2001, https://www.grc.nasa.gov/WWW/K-12/DLN/descriptions/docs/SBreallunar_landing.pdf.

4. Janie-Lynn Kang and Lisa M. Williamson, "Unemployment Rate Nears Prerecession Level by End of 2015," U.S. Bureau of Labor Statistics, April 2016, https://doi.org/10.21916/mlr.2016.19; Judith Stepan-Norris and Jasmine Kerrissey, "Over 300,000 Workers Have Gone on Strike This Year, but Don't Be Fooled—Unions Are Still a Dim Shadow of the Midcentury Glory Days," *Fortune,* August 24, 2023, https://fortune.com/2023/08/26/how-big-summer-of-strikes-labor-movement-union-membership-history/; U.S. Department of Energy, Office of Scientific and Technical Information, "International Petroleum Annual, 1975. [Events of 1976 and first part of 1977]," OSTI, accessed November 5, 2023, https://www.osti.gov/biblio/6860501.
5. For the 1975–81 timeline on presidential campaigns, I relied extensively on Rick Perlstein's book series documenting the phenomenon of the rise of the conservative movement in the 1970s.
6. "1976 Presidential Election: The Republican Primaries—January–July 1976," Gerald R. Ford Presidential and Library Museum, https://www.fordlibrarymuseum.gov/museum/exhibits/1976election/primaries.html, (accessed November 15, 2024).
7. Rick Perlstein, *The Invisible Bridge: The Fall of Nixon and the Rise of Reagan* (New York: Simon & Schuster, 2014), 731, 746.
8. Perlstein, *Invisible Bridge*, 779.
9. "A Time for Choosing Speech, October 27, 1964," Ronald Reagan Presidential Library and Museum, accessed November 3, 2023, https://www.reaganlibrary.gov/reagans/ronald-reagan/time-choosing-speech-october-27-1964.
10. "Republican National Convention Speech 1976," Ronald Reagan Presidential Library and Museum, accessed November 3, 2023, https://www.reaganlibrary.gov/archives/speech/republican-national-convention-speech-1976.
11. Ibid.
12. Ibid.
13. It wasn't until the 1980 presidential election cycle that conservative groups like Jerry Falwell's Moral Majority rallied behind the issue of abortion to help elect converted pro-life candidate, Ronald Reagan. Reagan's victory was also due to so-called Reagan Democrats—white, working-class Democrats—who were hit hard by the recession, turned off by social issues and turned on to Reagan's winning message of "making

America great again."

14. Judith Stein, *Pivotal Decade*, (New Haven: Yale University Press), 2010, 149.
15. "Time for Choosing Speech"; Ronald Reagan, "Excerpts of remarks by the Hon. Ronald Reagan before the Intercollegiate Studies Institute Banquet," Intercollegiate Studies Institute, Washington, DC, January 15, 1977, https://www.reaganlibrary.gov/public/2024-02/40-656-7386407-431-010-2023.pdf.
16. Rick Perlstein, *Reaganland: America's Right Turn, 1976–1980* (New York: Simon & Schuster, 2020), 67.
17. T. R. Reid, "Republicans Bask in Minnesota House Race," *Washington Post*, February 24, 1977, https://www.washingtonpost.com/archive/politics/1977/02/24/republicans-bask-in-win-in-minnesota-house-race/b2869369-1bd6-4943-ae21-679f9653a98d/.
18. Perlstein, *Reaganland*, 101.
19. Ronald Reagan, "Labor Day Address at Liberty Park," Jersey City, New Jersey, September 1, 1980, https://www.americanrhetoric.com/speeches/ronaldreaganlibertypark.htm.
20. Ronald Reagan, "Acceptance Speech," Republican National Convention, Detroit, Michigan, July 17, 1980, https://www.reaganlibrary.gov/public/2021-12/40-910-002-006-2021.pdf.
21. Jimmy Carter, "Anti-Inflation Policy Remarks to Members of the American Society of Newspaper Editors Announcing the Administration's Policy," Address to the American Society of Newspaper Editors, April 11, 1978, https://www.presidency.ucsb.edu/documents/anti-inflation-policy-remarks-members-the-american-society-newspaper-editors-announcing.
22. Stephanie Hugie Barello, "Consumer Spending and U.S. Employment from the 2007–2009 Recession through 2022," U.S. Bureau of Labor Statistics, October 2014, https://doi.org/10.21916/mlr.2014.34.
23. Perlstein, *Reaganland*, 204.
24. Perlstein, *Reaganland*, 204.
25. Perlstein, *Reaganland*, 230.
26. Douglas Matthews, Grant Ujifusa, and Michael Barone, *The Almanac of American Politics 1980* (New York: E. P. Dutton, 1979).
27. Jason Sisney et al., "Proposition 13 Report," The California Legislature's Nonpartisan Fiscal and Policy Advisor, September 22, 2016, https://lao.ca.gov/LAOEconTax/article/Detail/209; Chris Nichols, Taxpayers Revolt," Capital Public Radio, October 25, 2018, https://www.kpbs.org/news/midday-edition/2018/10/25/birth-californias-taxpayer-revolt.
28. Federal Election Commission, *FEC Releases Summaries on Non-Party Political Committee:*

1977–78 Financial Activity, (Washington, D.C.: Federal Election Commission, November 2, 1978).

29. William E. Farrell, "Begin and Other Israelis Deplore Senate Vote on Jets," *New York Times*, May 17, 1978.
30. Perlstein, *Reaganland*, 380–83.
31. Macomb County voted Republican in 1976, 52 percent to 46 percent. Reagan won by the same margin in 1980, 49 percent to 43 percent. However, the 1976 nominee, Gerald Ford, was a former Michigan congressman and likely benefited from a native son effect; "Recent Presidential Elections," 270 to Win, accessed September 5, 2024, https://www.270towin.com/states/Michigan.
32. Stanley B. Greenberg, "Middle Class Dreams," Democracy Corps, 1996, https://democracycorps.com/wp-content/uploads/1996/03/Middle-Class-Dreams-Chapter-2-Macomb-in-the-American-Mind.pdf.
33. James M. Hohman, "Michigan Adds Jobs as Auto Employment Plummets: A Chart to Ponder," Mackinac Center for Public Policy, December 2, 2022, https://www.mackinac.org/blog/2022/michigan-adds-jobs-as-auto-employment-plummets; Mark J. Perry, "Big Slide for Big 3 Market Share: From 90% to 45%," American Enterprise Institute, January 18, 2010, https://www.aei.org/carpe-diem/big-slide-for-big-3-market-share-from-90-to-45/; David Halberstam, *The Reckoning* (New York: William Morrow, 1986).
34. Greenberg, "Middle Class Dreams."
35. Stanley B. Greenberg, "Goodbye, Reagan Democrats," *New York Times*, November 10, 2008, https://www.nytimes.com/2008/11/11/opinion/11greenberg.html.
36. David A. Dulio and John S. Klemanski, "Opinion: Election Recap: How Macomb and Oakland Counties Are Shifting," *Detroit News*, November 19, 2020, https://www.detroitnews.com/story/opinion/2020/11/20/opinion-election-recap-how-macomb-and-oakland-counties-shifting/3777544001/.
37. U.S. Census Bureau, "Bachelor's Degree or Higher (5-year estimate) in Macomb County, MI," Federal Reserve Bank of St. Louis, updated December 7, 2023, https://fred.stlouisfed.org/series/HC01ESTVC1726099.
38. U.S. Census Bureau, "Bachelor's Degree or Higher (5-year estimate) in Oakland County, MI," Federal Reserve Bank of St. Louis, updated December 7, 2023, https://fred.stlouisfed.org/series/HC01ESTVC1726125.
39. Outagamie County Clerk, "Election Results—2020," https://www.outagamie.org/home/showpublisheddocument/78357/637400524891100000; State Elections Commission, "Election Results—2008," https://whs.access.preservica.com/uncategorized/SO_ae131b98-b41e-46fc-b38e-81340b3a543d/.

40. The states included in the study were Pennsylvania, Ohio, Michigan, Indiana, Illinois, Wisconsin, Iowa, Minnesota, and Missouri; Richard J. Martin, *A 10-State Analysis of the Democratic Presidential Vote Decline in Working-Class Counties, 2012–2020* (PowerPoint Slides), American Family Voices and 21st Century Democrats, https://www.americanfamilyvoices.org/_files/ugd/d4d64f_5c5b228a260e4793ad6034c9dd535d36.pdf.
41. Approximately seven communities in Brown and Outagamie counties—"new" bellwether counties—could be classified as "factory towns" under the study's definition.
42. "Reagan's Nashua Moment," TheMacallan223, YouTube, posted February 22, 2008, https://youtu.be/OO2_49TycdE.
43. Bill Peterson, "Bush Ends 2-Year Quest, Concedes '80 Republican Nomination to Reagan," *Washington Post*, May 27, 1980, https://www.washingtonpost.com/archive/politics/1980/05/27/bush-ends-2-year-quest-concedes-80-republican-nomination-to-reagan/cb34c445-b892-4fc6-a859-a83810509077/.
44. Kai Bird, *The Outlier: The Unfinished Presidency of Jimmy Carter* (New York: Crown, 2021), 592.
45. Stuart Eizenstat, *President Carter: The White House Years* (New York: St. Martin's Press. 2018), 8.
46. Eizenstat, *President Carter*, 837.
47. George Wallace, who ran as an independent for president in 1968, took another stab at the presidency in 1976, this time as a Democrat. Unlike Carter, he made little effort to appeal to voters beyond hardened conservatives who shared his right-wing world view. Consequently, he won just three primary contests.
48. David K. Henry and Richard P. Oliver, "The Defense Buildup, 1977–85; Effects on Production and Employment," *Monthly Labor Review*, August 1987, https://www.bls.gov/opub/mlr/1987/08/art1full.pdf.
49. Martin Tolchin, "House Approves Budget Plan Supported by Reagan, 270–154; Democrats Split on Key Vote/President 'Grateful,'" *New York Times*, May 8, 1981.
50. William Greider, "The Education of David Stockman," *Atlantic*, December 1981, https://www.theatlantic.com/magazine/archive/1981/12/the-education-of-david-stockman/305760/
51. Ed Kilgore, "Requiem for the DLC," *New Republic*, February 8, 2011, https://newrepublic.com/article/83086/democratic-leadership-council-requiem-dlc.
52. Walter Shapiro, "Baby-Boomer Bill Clinton: A Generation Takes Power," *TIME*, November 16, 1992, https://content.time.com/time/magazine/0,9263,7601921116,00.html.
53. David Thomas, "How Arkansas's Economy Benefits from International Trade and Investment" Business Roundtable, https://tradepartnership.com/wp-content/

uploads/2019/01/BRT_TRADE_2018_Arkansas.pdf; Nathan Bomey and Worth Sparkman, "Arkansas's Union Workers Charted," Axios, January 31, 2024, https://www.axios.com/local/nw-arkansas/2024/01/31/arkansas-union-workers-charted.

54. James Surowiecki, "Bonds and Domination," *New York Magazine*, March 1, 1999, https://nymag.com/nymetro/news/bizfinance/columns/bottomline/199/; Bob Woodward, *The Agenda: Inside the Clinton White House* (New York: Pocket Books, 1995), 84.
55. "Republican Presidential Candidates Seek Donald Trump Stamp of Approval," MassLive, September 26, 2011, https://www.masslive.com/politics/2011/09/republican_presidential_candid_2.html.
56. "How Donald Trump Perpetuated the 'Birther' Movement for Years," ABC News, September 16, 2016, https://abcnews.go.com/Politics/donald-trump-perpetuated-birther-movement-years/story?id=42138176.
57. Jeffrey M. Jones, "Voter Engagement Lower than in 2010 and 2006 Midterms," Gallup, October 8, 2014, https://news.gallup.com/poll/178130/voter-engagement-lower-2010-2006-midterms.aspx.
58. Domenico Montanaro, "Clinton has 45 to 1 'Superdelegate' Advantage over Sanders," NPR, November 13, 2015, https://www.npr.org/2015/11/13/455812702/clinton-has-45-to-1-superdelegate-advantage-over-sanders; Eamon Javers, "Bernie's Long Odds vs. Hillary: NBC-WSJ Poll," CNBC, June 23, 2015, https://www.cnbc.com/2015/06/22/g-odds-vs-hillary-nbcwsj-poll.html.
59. "What Donald Trump Has Said About Mexico and Vice Versa," ABC News, August 31, 2016, https://abcnews.go.com/Politics/donald-trump-mexico-vice-versa/story?id=41767704.
60. "2016 National Primary Polls" FiveThirtyEight, July 1, 2016, https://projects.fivethirtyeight.com/election-2016/national-primary-polls/republican/.
61. "Sanders in Dramatic Finish to Clinton Nomination," Roll Call, July 26, 2016, https://rollcall.com/2016/07/26/sanders-in-dramatic-finish-to-clinton-nomination/.
62. Altogether, Sanders received 42 percent of the Democratic primary vote and secured 1,893 delegates; "CNN Delegate Estimate," CNN, accessed September 5, 2024, https://www.cnn.com/election/2016/primaries/parties/democrat.
63. John Cassidy, "Piketty's Inequality Story in Six Charts," *New Yorker*, March 26, 2014.
64. "Student Debt Has Increased Sevenfold over the Last Couple Decades. Here's Why," Peter G. Peterson Foundation, October 26, 2021, https://www.pgpf.org/blog/2021/10/student-debt-has-increased-sevenfold-over-the-last-couple-decades-heres-why.
65. Bruce P. Frohnen, "American Populism in the Early Twenty-First Century: Constitutional Resistance to the New Class," *The Independent Review* 26, no. 1 (2021): 39–51, https://www.jstor.org/stable/48647334.

66. Edwin Warner, "The New Populism: Radicalizing the Middle," *TIME*, April 17, 1972, https://content.time.com/time/subscriber/article/0,33009,944465,00.html.
67. Robert Pear, "Health Law Repeal Could Cost 18 Million Their Insurance, Study Finds," *New York Times*, January 19, 2017, https://www.nytimes.com/2017/01/17/us/politics/congressional-budget-office-affordable-care-act.html.
68. Charles Postel, "If Trump and Sanders Are Both Populists, What Does Populist Mean?" Process: A Blog for American History, Organization of American Historians, accessed November 15, 2024, https://www.oah.org/process/if-trump-and-sanders-are-both-populists-what-does-populist-mean/.
69. Andrew Jackson, "Fifth Annual Message," Washington, DC, December 3, 1833, https://www.presidency.ucsb.edu/documents/fifth-annual-message-2.
70. Populist parties of the nineteenth and early twentieth centuries favored free trade and opposed protectionist policies because they viewed the latter as corporate subsidies at the expense of small, family farmers and workers.
71. Danielle Kurtzleben, "Here's How Many Bernie Sanders Supporters Ultimately Voted for Trump." NPR, August 24, 2017. https://www.npr.org/2017/08/24/545812242/1-in-10-sanders-primary-voters-ended-up-supporting-trump-survey-finds.
72. Philip Bump, "Donald Trump Will Be President Thanks to 80,000 People in Three States," *Washington Post*, December 1, 2016, https://www.washingtonpost.com/news/the-fix/wp/2016/12/01/donald-trump-will-be-president-thanks-to-80000-people-in-three-states/.
73. Andrew Glass, "Bush Declared Elector Victor Over Gore, December 12, 2000," *Politico*, December 12, 2018, https://www.politico.com/story/2018/12/12/scotus-declares-bush-electoral-victor-dec-12-2000-1054202.
74. Eizenstat, *President Carter*, 63.
75. Impropriety clouded the outcomes of the 1960 and 2000 election outcomes, the former marred by ballot stuffing in Texas and Illinois, and the latter by a largely partisan Supreme Court decision that ended recounts.
76. "ICYMI: President Biden: I'm the Most Pro-Union President in American History. And I Make No Apologies for It," AFL-CIO, June 29, 2023, https://aflcio.org/statements/icymi-president-biden-im-most-pro-union-president-american-history-and-i-make-no; AFSCME. "Saunders: 'President Biden Is Most Pro-Union, Pro-Worker President Of Our Lifetimes,'" April 25, 2023, https://www.afscme.org/blog/saunders-president-biden-is-most-pro-union-pro-worker-president-of-our-lifetimes.
77. "American Rescue Plan," The White House, accessed September 5, 2024, https://www.whitehouse.gov/american-rescue-plan/.

CHAPTER 8. THE RISE AND FALL OF THE AMERICAN SHIPBUILDING INDUSTRY

1. This is a rough translation of a phonetic, politically incorrect interpretation of the muskrat hunter's words with Pessano as recorded in *Meandering Michigan*'s "Little Ecorse and River Rouge Built and Launched Big Ships: The Story of the Great Lakes Engineering Works" (https://meanderingmichiganhistory.weebly.com/little-ecorse-and-river-rouge-built-big-ships—the-story-of-the-great-lakes-engineering-works.html).
2. Kathy Warnes, "Little Ecorse and River Rouge Built and Launched Big Ships: The Story of the Great Lakes Engineering Works," Meandering Michigan History, accessed October 5, 2023, https://meanderingmichiganhistory.weebly.com/little-ecorse-and-river-rouge-built-big-ships--the-story-of-the-great-lakes-engineering-works.html.
3. Skip Meier and Wayne Garrett, eds., *The Great Lakes Engineering Works: The Shipyard and Its Vessels*, (Detroit, MI: The Marine Historical Society, 2008), https://www.mhsd.org/glew/1shipyard.pdf.
4. Warnes, "Little Ecorse."
5. Meier and Garrett, *Great Lakes Engineering Works*.
6. Ibid.
7. Alden Whitman, "Walter Lippmann, Political Analyst, Dead at 85," *New York Times*, December 15, 1974, https://www.nytimes.com/1974/12/15/archives/walter-lippmann-political-analyst-dead-at-85-walter-lippmann.html.
8. It bears noting that many of the Democrats who overrode Truman's veto were southern Democrats, a more conservative and racist variant of their northern counterparts. The union profile of the workforce in the south was much smaller than in the north.
9. Basil Harris, E. S. Land, Lewis W. Douglas, and J. F. Gehan, "Ships and Subsidies," *Atlantic*, July 1945, https://www.theatlantic.com/magazine/archive/1945/07/ships-and-subsidies/656130/.
10. Ibid.
11. Department of Defense, Maritime Administration, Comptroller General of the United States, "Government Support of the Shipbuilding Industrial Base," Government Accountability Office, February 12, 1975, https://www.gao.gov/assets/psad-75-44.pdf.
12. Ibid.
13. "Construction Differential Subsidy," Justia, accessed September 6, 2024, https://www.justia.com/dictionary/construction-differential-subsidy/.
14. Loren Thompson, "U.S. Shipbuilding Is at Its Lowest Ebb Ever. How Did America Fall So Far?" *Forbes*, July 27, 2021, https://www.forbes.com/sites/lorenthompson/2021/07/23/us-shipbuilding-is-at-its-lowest-ebb-ever-how-did-america-fall-so-far/?sh=e6345dd6c87a.

15. Ajay Menon, "List of Major Shipbuilding Companies Around the Globe," Marine Insight, October 12, 2020, https://www.marineinsight.com/know-more/list-of-major-shipbuilding-companies-around-the-globe/.
16. "Ship Building in the US—Market Size, Industry Analysis, Trends and Forecasts (2023–2028)," IBISWorld, https://www.ibisworld.com/united-states/market-research-reports/ship-building-industry/#Methodology.
17. Jude Blanchette, Jonathan E. Hillman, Mingda Qiuand, and Maesea McCalpin, "Hidden Harbors: China's State-Backed Shipping Industry," Center for Strategies and International Studies, CSIS Briefs, July 8, 2020, https://www.csis.org/analysis/hidden-harbors-chinas-state-backed-shipping-industry.
18. Manoj Kumar, "Exclusive: India to Offer Incentives to Boost Shipbuilding Industry—Sources," Reuters, December 1, 2022, https://www.reuters.com/markets/asia/india-plans-offer-slew-incentives-boost-shipbuilding-industry-sources-2022-12-01/.
19. The Michigan legislature voted to repeal their right-to-work law. The new law will take effect on March 30, 2024.
20. Patrick Marley, "Act 10 at 10: Four Takeaways from the Scott Walker Legislation That Shook Wisconsin," *Milwaukee Journal-Sentinel*, February 8, 2021. https://www.jsonline.com/story/news/politics/2021/02/08/act-10-turns-10-four-takeaways-law-shook-wisconsin/4402917001/.
21. Lane Windham, *Knocking on Labor's Door: Union Organizing in the 1970s and the Roots of a New Economic Divide* (Chapel Hill: University of North Carolina Press, 2017), 1–3, 105.
22. Michael Fabey, *Heavy Metal: The Hard Days and Nights of the Shipyard Workers Who Build America's Supercarriers* (New York: William Morrow, 2022).
23. Ibid.
24. Megan Eckstein, "Updated: Navy Awards 2-Carrier Contract to Newport News Shipbuilding," USNI News, January 31, 2019, https://news.usni.org/2019/01/31/navy-awards-2-carrier-contract-newport-news-shipbuilding.
25. Ronald Reagan, inaugural address, January 20, 1981, https://www.reaganlibrary.gov/archives/speech/inaugural-address-1981.
26. Brent Sadler, "For the Larger Navy U.S. Needs, American Shipping and Shipbuilding Must Be Reviewed," The Heritage Foundation, October 27, 2022, https://www.heritage.org/defense/commentary/the-larger-navy-us-needs-american-shipping-and-shipbuilding-must-be-revived.
27. Jerry Hendrix, "The Navy Needs More Public Shipyards," National Review, February 5, 2023, https://www.nationalreview.com/2023/02/the-navy-needs-more-public-shipyards/.

CHAPTER 9. CHRISTMAS IN JULY

1. Laura R. Ashlee, "Broken in Two," *Michigan History* 74, no. 6 (Nov/Dec 1990): 32–37, as cited in The Presque Isle Historical Society, "The Sinking of the Carl D. Bradley," https://pichmuseum.org/carldbradley/.
2. Ric Mixter (dir.), *Expedition '94: To the Fitzgerald.* Flint, MI, 1995.
3. Lania Rocha, "Swartz Creek Residents, Shipwreck Divers Give Presentation on Edmund Fitzgerald," *Swartz Creek View*, November 10, 2022, https://swartzcreekview.mihomepaper.com/articles/swartz-creek-residents-shipwreck-divers-give-presentation-on-edmund-fitzgerald/.
4. Sean Ley, interview with author, Sault Ste Marie, Michigan, October 1, 2023.
5. Scott Purvis, phone interview with author, October 4, 2023.
6. Tom Farnquist, interview with author, Sault Ste Marie, Michigan, September 30, 2023.
7. Reporters Committee for Freedom of the Press, "New Law Prohibits Photographing Underwater Corpses," RCFP, October 20, 1997, https://www.rcfp.org/new-law-prohibits-photographing-underwater-corpses/.
8. Although copies of the photograph are in circulation and I obtained one, I have chosen not to reproduce the photograph out of respect for the family.
9. Mike TenEyck, phone interview with author, October 3, 2023.
10. Mixter, *Expedition '94.*
11. Scott Purvis, phone interview with author, October 4, 2023.
12. Tom Farnquist, interview with author, Sault Ste Marie, Michigan, September 30, 2023.
13. Ric Mixter, "Tattletale Sounds: The Edmund Fitzgerald Investigations," Bay County Historical Society, YouTube, posted January 18, 2023, https://youtu.be/OW-tig-2P-I.
14. Tom Farnquist, interview with author, Sault Ste Marie, Michigan, September 30, 2023.
15. Noah Bierman, "Harbor Branch Institute Becoming Part of Florida Atlantic University," *Herald-Tribune*, June 18, 2007, https://www.heraldtribune.com/story/news/2007/06/18/harbor-branch-institute-becoming-part-of-florida-atlantic-university/28566053007/.
16. Steve Thomas, "Subs Evoke Memories as Harbor Branch Celebrates 50th," VeroNews.com, April 17, 2022, https://veronews.com/2022/04/17/subs-evoke-memories-as-harbor-branch-celebrates-50th/.
17. Bierman, "Harbor Branch Institute."
18. Tom Farnquist, interview with author, Sault Ste Marie, Michigan, September 30, 2023.
19. "The Edmund Fitzgerald: Crew Probes Ship's Underwater Tomb," [Hazelton, PA] *Standard-Speaker*, July 6, 1994, https://www.newspapers.com/article/standard-speaker/1949390/.
20. It is unclear if Shannon's party was the first to find a body. In 1989 and again in 1994,

Farnquist made clear to his crew before each dive that if a body were found the cameras would be turned off and the discovery would not be reported. It is possible Farnquist had discovered a body (or more than one) long before Shannon.

21. Christina Sams, interview with author, Sault Ste Marie, Michigan, September 30, 2023.
22. I communicated with Janice Armagost several times, but Ms. Armagost was unwilling to go on the record.
23. Deborah Champeau, phone interview with author, August 11, 2023.
24. "Discovering the Titanic: Photo Gallery," *National Geographic*, October 29, 2009.
25. "Congressman Stupak's Edmund Fitzgerald Debate," bnodurft, YouTube, posted November 7, 2007, https://youtu.be/cutH-4Pzztc.
26. "President Approval Ratings—Bill Clinton," Gallup, accessed November 5, 2023, https://news.gallup.com/poll/116584/presidential-approval-ratings-bill-clinton.aspx.
27. Ann Devroy and John F. Harris, "The President is Relevant," *Washington Post*, April 19, 1995.
28. Tom Farnquist, interview with author, Sault Ste Marie, Michigan, September 30, 2023.
29. Mixter, "Tattletale Sounds."
30. "The Story of the Edmund Fitzgerald—Background," National Weather Service, accessed November 5, 2023, https://www.weather.gov/mqt/fitz_fitza.
31. YiLi Chien and Paul Morris, "Is US Manufacturing Really Declining?" Federal Reserve Bank of St. Louis, Our Economy Blog, April 11, 2017, https://www.stlouisfed.org/on-the-economy/2017/april/us-manufacturing-really-declining.
32. Heather Long, "NAFTA Is 'Killing the American Dream,' Michigan Workers Say," CNN Business, February 16, 2017, https://money.cnn.com/2017/02/16/news/economy/donald-trump-nafta/.
33. Davin Wiitanen, "History of the K.I. Sawyer Air Force Base," Military History of Upper Great Lakes, October 20, 2017, https://ss.sites.mtu.edu/mhugl/2017/10/20/history-of-the-k-i-sawyer-air-force-base/.
34. Mixter, "Tattletale Sounds."
35. Michael Schumacher, *Mighty Fitz: The Sinking of the Edmund Fitzgerald* (New York: Bloomsbury, 2005), 169–70.
36. "Emory Kristof," Annenberg Space for Photography, April 17, 2011, https://www.annenbergphotospace.org/person/emory-kristof/.
37. "In Memory of Phil Nuytten, Newtsuit Inventor," Great Lakes Shipwreck Museum, May 17, 2023, https://shipwreckmuseum.com/in-memory-of-phil-nuytten-newtsuit-inventor/.
38. Tom Farnquist, interview with author, Sault Ste Marie, Michigan, September 30, 2023.
39. James Haskell, interview with author, Sturgeon Bay, Wisconsin, July 6, 2023.

40. Ed Garsten, "Edmund Fitzgerald Bell Raising 1995," CNN report, YouTube, posted November 11, 2020, https://youtu.be/_zQStHJb_cQ.
41. "A Raised Bell Recalls the Sunken Fitzgerald," *New York Times*, July 9, 1995.
42. Tom Farnquist, interview with author, Sault Ste Marie, Michigan, September 30, 2023.
43. Ibid.

CHAPTER 10. HARD KNOCK LIFE

1. Aimée R. Dechter, "World War II Mobilization in Men's Work Lives: Continuity or Disruption for the Middle Class?" *American Journal of Sociology* 110, no. 3 (2004): 761, https://doi.org/10.1086/422662.
2. Fred Armagost, interview with author, Iron River, Wisconsin, June 30, 2023.
3. Mike Simonson, "Lake Superior Trials," Wisconsin Public Radio, 1994, https://americanarchive.org/catalog/cpb-aacip-526-154dn40s6r.
4. John O'Brien, phone interview with author, August 30, 2023.
5. Deborah Champeau, phone interview with author, August 11, 2023.
6. Nicquel Terry Ellis, "Most Major Metropolitan Areas Have Become More Racially Segregated, Study Shows," CNN, June 21, 2021, https://www.cnn.com/2021/06/21/us/housing-segregation-cities-berkeley-study/index.html.
7. Poorest: Barbara Miner, "A Closer Look at Milwaukee Zip Code 53206," *Milwaukee Magazine*, January 18, 2015, https://www.milwaukeemag.com/milwaukee-zip-code-53206/; wealthiest: "Highest Household Income ZIPs in Wisconsin," Unitedstateszipcodes.org, 2017, https://www.unitedstateszipcodes.org/rankings/zips-in-wi/median_household_income/.
8. Alyssa Bloechl, "Families of the Fitzgerald Remember," *Green Bay Press Gazette*, November 10, 2015, https://www.greenbaypressgazette.com/story/news/local/door-co/news/2015/11/10/families-fitzgerald-remember/75526204/.
9. Deborah Champeau, phone interview with author, August 11, 2023.
10. "Tragedy Ends," *Green Bay Press Gazette*, November 16, 1975.
11. Jack Metzgar, *Striking Steel: Solidarity Remembered* (Philadelphia: Temple University Press, 2000), 152.
12. David Maraniss, *First in His Class: A Biography of Bill Clinton* (New York: Simon & Schuster, 1996).
13. Jon Shelton, *The Education Myth* (Ithaca, NY: Cornell University Press, 2023).
14. John O'Brien, phone interview with author, August 30, 2023.

15. "The 100 Best Singles of 1990," *Slant Magazine*, January 10, 2011, https://www.slantmagazine.com/music/best-singles-of-the-1990s/.
16. James Haskell, interview with author, Sturgeon Bay, Wisconsin, July 6, 2023.

CHAPTER 11. GREAT NATIONS BUILD THINGS

1. Pulp and paper industry employment data computed by Ryan Long, regional economist for the Bay Area Workforce Development Board (Wisconsin), November 5, 2023.
2. For a full treatment of the subject, see Thomas M. Nelson, *One Day Stronger: How One Union Local Saved a Mill, Changed an Industry and What it Means for the Future of Manufacturing* (Irvington, NY: Rivertowns Books, 2021).
3. "SkyWater Thanks Congress for Passing the CHIPS and Science Act of 2022," (news release), Skywater Technology, July 28, 2022, https://www.skywatertechnology.com/skywater-thanks-congress-for-passing-the-chips-and-science-act-of-2022/.
4. Catherine Thorbecke, "The US Is Spending Billions to Boost Chip Manufacturing. Will It Be Enough?" CNN, October 18, 2022, https://www.cnn.com/2022/10/18/tech/us-chip-manufacturing-semiconductors/index.html.
5. Troy Segal, "Troubled Asset Relief Program, What It Was, How It Worked," Investopedia, September 7, 2023, https://www.investopedia.com/terms/t/troubled-asset-relief-program-tarp.asp.
6. "Top Contributors to Political Parties," NPR, March 2, 2011, https://www.npr.org/2011/03/03/134209755/top-contributors-to-political-parties.
7. "The Auto Bailout at 10 Years Later: Was It the Right Call?" Knowledge at Wharton, September 12, 2018, https://knowledge.wharton.upenn.edu/podcast/knowledge-at-wharton-podcast/auto-bailout-ten-years-later-right-call/.
8. Monali Samaddar and Daniel Bachman, "No Longer a Smooth Drive: How Automobiles' Role in the US Economy Has Evolved," Deloitte Insights, February 2022, https://www2.deloitte.com/us/en/insights/economy/spotlight/automobile-impact-us-economy.html.
9. Peter Behr, "When the Taxpayers Saved Lee Iacocca's Bacon, What Did They Get Out of It?" *The Washington Post*, May 13, 1984.
10. "Episode Guide: Hell on Wheels," IMDb, https://www.imdb.com/title/tt1699748/.
11. For an overview of these laws and their economic impact, see Nelson, *One Day Stronger*.
12. Kathleen Elkins, "A Brief History of the 401(k), Which Changed How Americans Retire," CNBC, January 4, 2017, https://www.cnbc.com/2017/01/04/a-brief-history-of-the-401k-which-changed-how-americans-retire.html.

13. Paul N. Van De Water and Kathleen Romig, "Social Security Benefits Are Modest," Center on Budget and Policy Priorities, January 8, 2020, https://www.cbpp.org/research/social-security/social-security-benefits-are-modest.
14. Gretchen Morgenson and Joshua Rosner, *These Are the Plunderers: How Private Runs—and Wrecks—America* (New York: Simon & Schuster, 2023).
15. Marc Sandalow, "Bush Claims Mandate, Sets 2nd-Term Goals," SFGATE, November 5, 2004, https://www.sfgate.com/politics/article/Bush-claims-mandate-sets-2nd-term-goals-I-2637116.php.
16. Tami Luhby, "Social Security Benefits Have Lost 36% of Buying Power Since 2000," CNN, May 10, 2023, https://www.cnn.com/2023/05/10/politics/social-security-benefits-inflation/index.html.
17. National Institute on Retirement Security, "New Report: 40% of Older Americans Rely Solely on Social Security for Retirement Income" (news release), NIRS, January 13, 2020, https://www.nirsonline.org/2020/01/new-report-40-of-older-americans-rely-solely-on-social-security-for-retirement-income/.
18. AP-NORC Center for Public Affairs Research, "Younger generations are not confident that Social Security or Medicare will be there for them," APNORC, April 2023, https://apnorc.org/projects/ayounger-generations-are-not-confident-that-social-security-or-medicare-will-be-there-for-them-a/.
19. Jennifer Dixon, "Poll Finds Young Americans Doubt Social Security Future," *Washington Post,* September 27, 1994, https://www.washingtonpost.com/archive/business/1994/09/27/poll-finds-young-americans-doubt-social-security-future/9d8df6b0-f0d6-4b35-8387-56a20cc8511b/.
20. Government of Canada, "National Shipbuilding Strategy," Public Services and Procurement Canada, April 4, 2023, https://www.tpsgc-pwgsc.gc.ca/app-acq/amd-dp/mer-sea/sncn-nss/apropos-about-eng.html.
21. T. Ring, "The National Shipbuilding Procurement Strategy: How Did We Get to Where We Are Now?" Canadian Global Affairs Institute, March 2016, https://www.cgai.ca/the_national_shipbuilding_procurement_strategy.
22. "Shipbuilding in the US: Market Size, Industry Analysis, Trends and Forecasts (2024–2029)," IBISWorld, January 2023, https://www.ibisworld.com/united-states/market-research-reports/ship-building-industry/#KeyStatistics.
23. Government of Canada, "National Shipbuilding Strategy."
24. Government of Canada, "About the National Shipbuilding Strategy," https://www.canada.ca/en/public-services-procurement/services/acquisitions/defence-marine/national-shipbuilding-strategy/about.html.

25. J. L. Granastein, “That Sinking Feeling—Despite a Huge Investment in Shipbuilding, Canada’s Navy Isn’t Living Up to Its Historic Reputation,” *Legion Magazine*, February 17, 2023, https://legionmagazine.com/that-sinking-feeling-despite-a-huge-investment-in-shipbuilding-canadas-navy-isnt-living-up-to-its-historic-reputation/.
26. Ryan Koronowski, Jessica Vela, Zahir Rasheed, and Seth Hanlon, “These 19 Fortune 100 Companies Paid Next to Nothing—or Nothing at All—in Taxes in 2021,” American Progress, April 26, 2022, https://www.americanprogress.org/article/these-19-fortune-100-companies-paid-next-to-nothing-or-nothing-at-all-in-taxes-in-2021/.
27. Kurt Van Dender, Alexander Pick, and Leonie Cedano, “Taxing Wages—The United States,” OECD, accessed November 5, 2023, https://www.oecd.org/tax/tax-policy/taxing-wages-united-states.pdf.
28. John O’Brien, phone interview with author, August 30, 2023.
29. Aldo Leopold, *A Sand County Almanac and Sketches Here and There*, 2nd ed. (Oxford: Oxford University Press, 1968).
30. For a complete analysis and historical account of this environmental phenomenon, see Gregory Summers, *Consuming Nature* (Lawrence: University of Kansas Press, 2006).
31. For additional information, see Nelson, *One Day Stronger*, chap. 4.
32. Tyler Platz, “Linking Variations in Wisconsin’s Cancer Incidence Rates and Environmental Carcinogens Present in the Lower Fox River Valley” (dissertation, St. Norbert College, 2013), https://schneiderschool.snc.edu/researchandservices/cbea/docs/cbeapapers/the_fox_river_final_paper.pdf.
33. National Cancer Institute, “Incidence Rates Table,” State Cancer Profiles, accessed November 3, 2023, https://www.statecancerprofiles.cancer.gov/incidencerates/.
34. John O’Brien, phone interview with author, August 30, 2023.
35. Sean Ley, interview with author, Sault Ste Marie, Michigan, October 1, 2023.
36. Deborah Champeau, phone interview with author, August 11, 2023.
37. Tom Temin, “If It Does Get Passed, the Defense Authorization Bill Has Quite a Few Surprises,” Federal News Network, October 17, 2023, https://federalnewsnetwork.com/defense-main/2023/10/if-it-does-get-passed-the-defense-authorization-bill-has-quite-a-few-surprises/.
38. Drew Desilver, “Congress Has Long Struggled to Pass Spending Bills,” Pew Research Center, September 13, 2023, https://www.pewresearch.org/short-reads/2023/09/13/congress-has-long-struggled-to-pass-spending-bills-on-time/.
39. “10 Largest Budget Functions,” Peter G. Peterson Foundation, May 24, 2023, https://www.pgpf.org/budget-basics/top-10-largest-budget-functions.
40. Alyssa Bloechl, “Families of the Fitzgerald Remember,” *Green Bay Press Gazette*,

November 10, 2015, https://www.greenbaypressgazette.com/story/news/local/door-co/news/2015/11/10/families-fitzgerald-remember/75526204/.

41. Aerial View of Manitowoc Shipbuilding Company Shipyards" (photograph), University of Wisconsin–Madison Libraries, 1943, https://search.library.wisc.edu/digital/A2BTF4BD73W3CT9C.
42. Wisconsin Maritime Museum at Manitowoc, "USS Cobia & Manitowoc Submarines," accessed November 5, 2023, https://www.wisconsinmaritime.org/submarines/.

Index

A

B